RAND

02/19/02

MR. LENOIR –

I ATTACH RAND'S WORK ON BIOMETRICS ALONG W/ INFORMATION ON THE CTST 2002 CONFERENCE IN NEW ORLEANS.

CTST 2002 IS EXPENSIVE BUT ONE OF THE BEST BIOMETRIC EVENTS. PLEASE CONTACT ME W/ ANY QUESTIONS. John

John D. Woodward, Jr., Esq

ARMY BIOMETRIC APPLICATIONS

Identifying and Addressing Sociocultural Concerns

John D. Woodward, Jr., Katharine W. Webb, Elaine M. Newton, Melissa Bradley, David Rubenson

with Kristina Larson, Jacob Lilly, Katie Smythe, Brian Houghton, Harold A. Pincus, Jonathan M. Schachter, Paul Steinberg

Prepared for the United States Army
Approved for public release; distribution unlimited

Arroyo Center
RAND

The research described in this report was sponsored by the United States Army under Contract No. DASW01-96-C-0004.

ISBN 0-8330-2985-1

Cover design by Maritta Tapanainen

Published 2001 by RAND
1700 Main Street, P.O. Box 2138, Santa Monica, CA 90407-2138
1200 South Hayes Street, Arlington, VA 22202-5050
201 North Craig Street, Suite 102, Pittsburgh, PA 15213-1516
RAND URL: http://www.rand.org/
To order RAND documents or to obtain additional information, contact Distribution Services: Telephone: (310) 451-7002; Fax: (310) 451-6915; Internet: order@rand.org

PREFACE

The digitized Army of the twenty-first century depends on secure command, control, communications, and computers to ensure dominance on the battlefield. Biometrics has been suggested as a means to enhance this security.

This report documents RAND's findings regarding Army use of biometrics. The concerns raised and corresponding solutions will likely affect almost any organization intending to make use of biometrics and should be of interest to anyone concerned about the functions of these organizations.

Lieutenant General William H. Campbell, Director of Information Systems for Command, Control, Communications, and Computers (DISC4) and the Army's Chief Information Officer, sponsored this work. Phillip Loranger of the Information Assurance Office served as RAND's primary point of contact. The research was conducted in the Force Development and Technology Program of RAND's Arroyo Center, a federally funded research and development center sponsored by the U.S. Army.

For more information on RAND Arroyo Center, contact the Director of Operations (telephone 310-393-0411, extension 6500; FAX 310-451-6952; e-mail donnab@rand.org), or visit the Arroyo Center's Web site at http://www.rand.org/organization/ard/.

CONTENTS

FIGURES

TABLES

SUMMARY

INTRODUCTION

The U.S. Army has a growing need to control access to its systems in times of both war and peace. In wartime, the Army's dependence on information as a tactical and strategic asset requires the Army to carefully control its battlefield networks. From logistics flows to intelligence on enemy forces, the Army depends on confining access to its data to authorized personnel. This need for access control is also critical at the weapon system level.

Access control issues are important to the peacetime Army because improving the efficiency of peacetime operations, including controlling access to facilities, computer systems, and classified information, depends on fast and accurate identification. The Army also operates a vast set of human resource services involving health care, retiree and dependent benefits, and troop support services. These services create the need for positive identification to prevent fraud and abuse.

The use of biometrics has been proposed as a solution to these many needs. Biometrics are physical characteristics or personal traits of a person that can be measured and used to recognize that person either by identification or verification. Identification occurs when the biometric system identifies a person from the entire enrolled population by searching a database for a match. This process is sometimes called "one-to-many" matching. Verification occurs when the biometric system authenticates a person's claimed identity from his previously enrolled pattern. This is called "one-to-one" matching.

The potential of biometrics, combined with increased policymaker interest, has led the Army to undertake an intense assessment of biometric technologies. The Army is studying how it can use biometric applications to improve security, efficiency, and convenience, as well as whether it should establish an Army biometric center that could serve as a central data repository for biometric information and perform research, development, test, and evaluation (RDT&E) functions. Because interest in biometrics in the federal government is widespread, the Army is also examining the role a center could play in supporting a national biometrics program.

At the direction of Lieutenant General William H. Campbell, Director of Information Systems, Command, Control, Communications, and Computers (DISC4) and the Army's Chief Information Officer, RAND examined the legal, sociological, and ethical issues associated with the U.S. Army's use of biometrics and the establishment of an Army biometric center. RAND assembled an interdisciplinary team of researchers who reviewed literature and interviewed technologists, program managers, lawyers, ethicists, and privacy experts to identify issues and methods to address them. To test its conclusions, RAND conducted a workshop with a number of experts in mid-December 1999. The research, which focused on the United States, had the following four objectives:

- Provide an overview of biometric technologies.
- Identify sociocultural (meaning sociological, legal, and ethical) concerns that might be raised by Army use of biometrics and suggest solutions to mitigate these concerns.
- Analyze the feasibility of an Army biometric program, including a national biometric center and national data repository.
- Provide implementation recommendations for the Army, including suggested areas for further research.

OVERVIEW OF BIOMETRIC TECHNOLOGIES

In the context of this report, "biometrics" refer to commercially viable automated methods of identifying, or verifying the identity of, a living person in real time based on a physical characteristic or personal trait of the individual. This is commonly done by comparing a stored template (defined as a type of file record of a characteristic or trait) against a template of the live image captured through a sensor. While many possible biometrics exist, at least eight mainstream biometric authentication technologies have been deployed or pilot-tested in commercial applications in the public and private sectors. They are fingerprint, hand/finger geometry, facial recognition, voice recognition, iris scan, retinal scan, dynamic signature verification, and keystroke dynamics.

RAND researchers compared these eight mainstream biometrics by interviewing technologists, vendors, and program managers, as well as studying the technical literature. They discovered that the utility and effectiveness of a biometric depends largely on the specific purposes for which it is used. Biometrics also vary widely in terms of intrusiveness, robustness, and distinctiveness.

WHAT SOCIOCULTURAL CONCERNS ARE RAISED BY USING BIOMETRICS?

As with all identification techniques, biometrics carry the potential to reduce the anonymity of our actions. In the United States, privacy has a great deal of value for our society and culture, and this importance is reflected in our laws. Hence, a feasibility assessment of Army use of biometrics and the establishment of a biometrics center must take into account the sociocultural issues that such use might raise.

We identified three major categories of concerns associated with biometrics: informational privacy, physical privacy, and religious objections.

Informational Privacy

Informational privacy, or an individual's ability to control information about himself, dominated the concerns of the experts interviewed. Specific informational privacy issues include

- function creep,
- tracking of individuals' activities, and,

- misuse of data, including identity theft.

Function Creep, in the context of biometrics, means that biometric data originally collected for one purpose are used for other purposes. Although using data for other secondary purposes might be worthwhile, sociocultural issues arise when individuals are not informed of these new purposes and have not given their consent to the new use.

Tracking refers to the specific function creep involved when biometrics are used to monitor an individual's actions or to search databases containing information about these actions. If a person must use the same standardized biometric to participate in life's everyday activities, he leaves a detailed record behind. This concern raises the question of whether using biometrics increases the ability to track individuals, possibly without their knowledge or consent.

Misuse of Information, in the form of data representing an individual's biometric, is also a potential problem. For example, using a biometric identifier, much the way a Social Security number (SSN) is used, to link a person's medical information with financial data, raises concerns. Misuse also includes concerns about identity theft. Because they are unique identifiers, biometrics should make identity theft more difficult; nonetheless, biometric data can be stolen or copied when used in certain ways.

Physical Privacy

Physical privacy concerns include

- stigmatization,
- actual harm, and,
- hygiene.

Stigmatization refers to the perception that biometrics carries a stigma. For example, fingerprinting has a strong association with criminal activity. These perceptions vary widely across cultures.

Actual Harm refers to the fear of some individuals that biometric technologies will actually do them physical harm. We have found no evidence that biometrics actually cause physical harm.

Hygiene refers to the fact that some object to using biometric devices that require touching a surface, for example, fingerprinting and hand geometry, because doing so might transmit germs from other individuals.

Religious Objections

Religious objections have been raised by certain Christian sects based on the "Mark of the Beast" language in the Book of Revelation. Although the number of these dissenters is small, some members of the Army community may hold similar beliefs. Thus, the Army must be prepared to address such objections.

ARE SUCH SOCIOCULTURAL CONCERNS NEW IN DEALING WITH BIOMETRICS?

Sociocultural concerns about biometrics are both similar and dissimilar to existing concerns regarding other government data collection and management efforts. Moreover, many of these same sociocultural concerns apply to private sector use of biometrics. Function creep, tracking, misuse of information, and identity theft are all longstanding concerns about the collection and storage of personal information. The use of the SSN illustrates the problem of function creep. When the Social Security Act was passed in 1935, promises were made to the American public that the SSN would never be used beyond its stated purpose—that is, to administer social security assistance. Today, despite these early assurances, an individual's SSN is used for many purposes, both in the public and private sectors.

On the other hand, despite these similarities, biometrics raise some concerns not associated with traditional identifiers, such as a password. Because biometrics are limited in number—humans have one face, 10 fingers, two eyes—concerns arise that if the biometric identification information is stolen, the individual would be unable to replace the identifier, which is relatively simple to do with today's current personal identification numbers (PINs). In addition, biometric data might contain medical information or indicate changes in medical conditions. While this is not the case for biometrics in use today, the potential is worrisome, because it would change the information available to organizations using biometrics as well as expand possibilities for misuse. Also, the capability to track individuals is significantly greater with biometrics than with traditional forms of identification. This capability may generate its own demand for use, leading to function creep. For these and other reasons, biometrics could be perceived by some as a qualitatively different means of checking identity.

HOW CAN THE ARMY MITIGATE THESE SOCIOCULTURAL CONCERNS?

The Army can rely on existing policies and procedures to address many of these concerns. However, because biometrics involve new technologies and our society increasingly focuses on the impact of information technology on privacy, it would be prudent for the Army to take a broad and integrated approach to managing its response to these sociocultural concerns.

Relying on Existing Laws and Regulations

Among the laws and regulations concerning government use of personal information, the Privacy Act of 1974 is most prominent. The Privacy Act regulates the collection, maintenance, use, and dissemination of personal information by federal agencies, including the Department of Defense (DoD) and the Army. Among its provisions, the Act addresses individual concerns related to personal information provided to a government agency. For example, the agency must state the purpose for collecting the data, its intended use of the information, and its authority to collect the information.

As a general rule, the Privacy Act prohibits a federal agency from disclosing personal information without the consent of the individual providing the information. However, the Act contains many exceptions to this rule. While the Privacy Act specifies the legal minimum the Army must do to be in compliance, the Army might want to provide broader privacy protections for its biometrics program. For example, it could take the position that no biometric data in its charge would be shared, similar to the rule DoD has for protecting DNA samples in its human remains identification program.

The Privacy Act also requires federal agencies and officials to protect their databases from unwarranted disclosures. This requirement addresses some of the concerns about misuse of data and identity theft.

The Army has regulations in place to accommodate religious objections. These regulations could be used to address religious objections to the use of biometrics.

In a case that has important implications for the Army, the U.S. Supreme Court has addressed informational privacy from a constitutional perspective. In *Whalen v. Roe*, the Court upheld a New York state law establishing a centralized computer database in which the state recorded and stored the names and addresses of all persons who obtained certain drugs pursuant to a doctor's prescription. The Supreme Court explained that New York had demonstrated its need for the database as part of its war on drugs and had taken extensive measures to prevent unauthorized disclosure of the data.

Similarly, the judiciary has addressed privacy concerns in related contexts. For example, the courts have consistently upheld federal, state, and local requirements for fingerprinting for employment and licensing, provided a rational basis existed for the requirement. Likewise, when a biometric is needed from an individual for a criminal justice purpose, the Army should be able to satisfy the constitutional requirements.

Taking a Broader, More Integrated Approach

No significant legal obstacles to Army use of biometrics in the United States have been identified. While the Army could rely on existing laws and regulations to provide a minimum level of privacy protection, the Army should take additional steps to strengthen privacy protections because it is in its best interest to do so. These additional steps include taking a broader, more integrated approach to mitigating sociocultural concerns. Such an approach includes four elements.

Step One: Thoroughly Explain Why Biometrics Are the Best Solution to a Particular Problem. This step requires a detailed statement of the problem, a description and evaluation of possible solutions, and a comparison of biometric capabilities to those of other potential solutions. This analysis will form the basis for individual and societal decisionmaking, balancing the benefits of biometrics against potential losses of privacy.

Step Two: Structure a Program and Select Technologies to Minimize the Effects on Privacy. This step will help prevent privacy concerns from arising. Within the constraints of meeting operational needs, Army decisionmakers should also consider the following:

- *Policies about sharing data* should be carefully designed to avoid perceptions of function creep and the development of tracking capability.
- *Privacy enhancing solutions* should be considered when the Army chooses biometric technologies. Specific examples that may allay privacy concerns about tracking include decentralizing template storage and matching; using nonforensic biometrics; using multiple biometrics; and using verification rather than identification applications. In addition, biometrics that are less intrusive or provide no medical information would likely be preferred by those concerned about privacy.
- *Data repository* choices affect perceptions of security and privacy. Holding template data on a smart card in the possession of the individual or locally with the sensor, rather than in a central repository, makes function creep and tracking less feasible.

Step Three: Educate the Army Community and the Public About the Purpose and Structure of the Program. Such an education program should explain what steps the Army has taken to ensure that privacy is protected. A campaign directed at both the Army community and the general public could generate support for the Army's program. The following questions should be addressed:

- What is the purpose of the biometric program? Who is included in it?
- What information will be available through the biometric?
- How will that information be used and who will have access to it?
- How will that information be protected?

- Who will establish, control, and review these practices?

Step Four: Assign Responsibility Within the Army for Guiding Steps One to Three. This step is important to ensure that sociocultural concerns are adequately addressed as the program is developed and implemented and as issues arise in the future.

WHAT IS THE FEASIBILITY OF AN ARMY OR NATIONAL BIOMETRIC CENTER?

Establishing an Army biometrics center, including both an RDT&E center and a central repository, has a good chance of success if it is based on justifiable program needs and structured to provide meaningful privacy protection. While legal, regulatory, technical, operational, security, and administrative issues affect how such a center could be established and what it could do, these do not impose overwhelming obstacles. However, a center may raise sociocultural concerns about privacy, with particular attention focused on a central repository. Concerns about a repository are likely to be much more sensitive to size, to purpose, and to who is in charge. While establishing a central repository could be justified based on particular purposes, it does not seem critical to the RDT&E effort. A centralized repository could help a test center verify the uniqueness of particular biometrics and algorithms by giving it large numbers of templates to compare, but this is only one of the activities that might be performed at an RDT&E center. In addition, biometric data are electronic records and could be sent relatively easily from one location to another. Thus, in this section, we address the RDT&E center and repository separately.

An *RDT&E center* could be justified by the need to focus activities in areas of interest to the government and to provide a forum to share information and coordinate activities across a number of organizations. While field or pilot testing and scenario evaluations are costly, they are the only reasonable methods to test a biometric system fully and reliably for deployment. Laboratory testing could be used to test algorithms and as an initial pass/fail test for biometric devices to achieve minimum standards for additional operational testing. An R&D lab could also undertake further development of mathematical and statistical methods for test design and evaluation of biometric systems. An R&D center could be a source of advice on biometric systems for agencies internal and external to the Army. It could advise other agencies regarding technology considerations and help them develop educational roll-out pieces for their biometric programs. Whether the Army would seek the role of center coordinator depends on the importance biometrics are expected to have in the Army. Whether this should be a truly national center or an Army or DoD center depends on the importance of biometrics to the nation at large.

A *central repository* will raise more sociocultural concerns. The explanation about why such a repository is needed should adhere closely to the points raised earlier about defining the purpose—explaining what the data will be use for, what additional data will be stored with the templates, who can access the data, and how data will be protected—and deciding who oversees these processes. Concerns about the repository are likely to depend on whose biometric identification information is included (e.g., only service members or also Department of the Army civilians, contractors, retirees, dependents, and foreign nationals).

A central repository could be justified by the need to use an identification biometric for the Army rather than simply relying on verification. Or, it may be necessary to have a centralized verification location so that certain identifiers can be used in multiple locations. A centralized repository could also help a test center verify the uniqueness of particular biometrics and algorithms, giving it large numbers of templates to compare.

A *national biometric center* must be justified in the same way by those most interested in having such a center. It is not clear that those most interested in RDT&E will also be most interested in a repository. While the military might want to move the technology forward, perhaps law enforcement or social service agencies will have the greatest interest in establishing some form of national repository for biometric data. These other agencies are probably most interested in comparing data to search for fraud or criminal evidence, activities likely to meet with sociocultural objections. An Army-run national biometric repository might not have such strong interests in sharing data with other agencies, although it would likely be under pressure to do so.

CONCLUSIONS AND RECOMMENDATIONS

Based on our analysis, we have identified no significant legal obstacles preventing the Army from establishing a biometrics program in the United States. Although some sociocultural concerns may arise, particularly with regard to privacy issues, these can be addressed, albeit minimally, by existing Army regulations, particularly those relating to the Privacy Act. To demonstrate its commitment to privacy, the Army should consider providing additional protection for its biometric databases beyond the requirements of the Privacy Act. In particular, the Army might want to place strict requirements on sharing biometric data with other agencies and organizations. If other agencies believe they have a legitimate claim to access to the Army's data, it might be better for Congress or the White House to decide this issue through the political process.

The Army should provide a detailed analysis of the problems that biometrics can help solve. This should include a detailed description of the problem, a description and evaluation of possible solutions, and a comparison of biometric capabilities to those of other potential solutions. This analysis will form the basis for individual and societal decisions balancing the benefits of biometrics against potential losses of privacy.

Although a central repository may be necessary, it should be justified in the same way as the Army biometric program, by establishing the need for a center based on specific problems to be addressed. The size and functions of this center will contribute to public perceptions and concerns about its purposes and potential threats to individual privacy.

Carefully targeted research could help the Army address sociocultural concerns when implementing its biometrics program. Greater Army participation in the U.S. government's Biometric Consortium could also assist Army and DoD research interests. As the Army uses biometrics overseas, it must consider international law issues. Moreover, the Army could benefit from research evaluating whether biometric data implicate medical information of any kind.

As this report was being prepared for final publication, Deputy Secretary of Defense Rudy de Leon issued a memorandum on December 27, 2000, consolidating oversight and management of biometric technology under the recently created DoD Biometrics Management Office (BMO). This memorandum also called for the formal establishment of a DoD Biometrics Fusion Center (BFC) under the BMO. The BFC's purpose is to acquire, test, evaluate, and integrate biometrics and to develop and implement storage methods for biometrics templates. The BFC is located in Bridgeport, West Virginia.

This memorandum derived from Public Law 106-246, signed by President Clinton on July 13, 2000, which included the following provision: "To ensure the availability of biometrics technologies in the Department of Defense, the Secretary of the Army shall be the Executive Agent to lead, consolidate, and coordinate all biometrics information assurance programs of the Department of Defense."[1]

As the DoD BMO and the Army, as executive agent, continue to assess biometrics, they must carefully consider the sociocultural concerns biometrics raise, along with technical, operational, security, bureaucratic, and administrative issues.

---

[1] For more information about the DoD Biometrics Management Office, please visit the DoD BMO Web page, available at http://www.c3i.osd.mil/biometrics/.

ACKNOWLEDGMENTS

The RAND biometrics team owes much to the many people who helped us with this report, from those who tutored us on the technologies to those who participated in the workshop, and all who gave generously of their time to answer our questions and provide insights. In particular, we thank Steve Goldberg of Georgetown University Law Center, Peter T. Higgins of Higgins and Associates, David Mussington of RAND, and James L. Wayman of the National Biometric Test Center for their comments on various aspects of this draft; Kristina Larson and Shirley C. Woodward for their thoughtful contributions to our effort; our colleagues, Elisa Eiseman and Priscilla Schlegel of RAND, for their expertise and timely inputs; Dan Sheehan of RAND, for his careful edits; and Nykolle Brooks and Jessica Boyd of RAND, for their help in putting this document together. Any errors, of course, are the responsibility of the authors alone.

ACRONYMS

AAFES	Army and Air Force Exchange Service
AFDC	Aid to Families with Dependent Children
AFIRM	Automated Fingerprint Image Reporting and Match
AFIS	Automated Finger Imaging system
BC	Biometric Consortium
BFC	Biometrics Fusion Center
BIOS	Basic input/output system
BMO	Biometrics Management Office
BPP	Bankruptcy petition preparer
CJIS	Criminal Justice Information Services (Division)
DCSPERS	Deputy Chief of Staff for Personnel (Army)
DEERS	Defense Enrollment Eligibility Reporting System
DFAS	Defense Finance and Accounting Service
DHS	Department of Human Services
DISC4	Director of Information Systems, Command, Control, Communications, and Computers
DMDC	Defense Manpower Data Center
DMV	Department of Motor Vehicles
DoD	Department of Defense
DPSS	Department of Public Social Services (Los Angeles)

EBT	Electronic benefit transfer
FMR	False match rate
FNMR	False nonmatch rate
FOIA	Freedom of Information Act
GAO	General Accounting Office
GSA	General Services Administration
IAFIS	Integrated Automated Fingerprint Identification System
INS	Immigration and Naturalization Service
INSPASS	INS Passenger Accelerated Service System
MISC	Military Identification Smart Card
NAC	National Agency Check
NBTC	National Biometric Test Center
NCIS	Naval Criminal Investigative Service
NIST	National Institute of Standards and Technology
NPL	National Physical Laboratory
NSA	National Security Agency
OGC	Office of the General Counsel (FBI)
OMB	Office of Management and Budget
PCMCIA	Personal Computer Memory Card International Association
PIN	Personal identification number
RDT&E	Research, development, test, and evaluation
RFID	Radio frequency identification device
SLC	Special Latent Cognizant
SSN	Social Security number
TANF	Temporary Assistance for Needy Families
TFA	Temporary Family Assistance
TRADOC	Training and Doctrine Command (Army)
UCMJ	Uniform Code of Military Justice

Chapter One

INTRODUCTION

BACKGROUND

The U.S. Army has a growing need to improve access control for its many systems, both in wartime and in peacetime. In wartime, the Army's dependence on information as a tactical and strategic asset requires it to carefully control its battlefield networks. From information on logistics flows to intelligence on enemy forces, the Army depends on confining access to its data to authorized personnel.

Access control is critical for weapon systems. These systems increasingly consist of physical, logical (computer), and informational components. Army weapon systems are so powerful and often so dominant that unauthorized use of even a single system can have significant adverse consequences. Moreover, if an enemy were to capture an Army weapon system with inadequate access control measures in place, the enemy could use the captured resource to its advantage.

In peacetime, access control issues are also important because improving the effectiveness and efficiency of Army operations depends on fast and accurate identification of authorized users. Examples include controlling access to facilities, computer systems, and classified information. Moreover, the Army operates a vast set of human resource services, involving health care, retiree and dependent benefits, troop support services, and many others. Access control is important in these systems to verify claims for benefits and to reduce fraud.

Biometrics is a possible solution for dealing with the Army's access control problems. Biometrics use distinctive physical characteristics or personal traits, such as fingerprints or hand geometry, to make nearly instantaneous verifications of claimed identity or to identify individuals. The push for biometrics is driven by both its technical possibilities and political interest.

From a technical point of view, commercially viable biometric authentication systems are in full-scale operation. Significant optimism has been expressed that technological improvements will lead to better, faster, less costly, and more pervasive systems. Because biometrics are an integral part of human beings or "bar codes for the body," they offer a convenience and efficiency that other identifiers, which must be remembered or produced, do not. For this reason, biometrics are seen as a means of enhancing security for activities currently protected by traditional means of access control—cards, personal identification numbers (PINs), and passwords.

Biometrics can also be used in conjunction with cards and PINs to enhance security. In many applications, a biometric could replace the card or PIN entirely. If there are no cards to lose or numbers to remember, in many cases biometrics will reduce operational and administrative costs and increase user convenience.

Some experts contend that biometrics, if properly used, could enhance privacy, along with security and convenience, by allowing for an individual's identity to be secured by different biometrics.[1] In other words, the use of multiple biometrics is the equivalent of an individual being issued multiple PINs or passwords, with the critical difference being that biometric-based systems provide better security and greater convenience. From the privacy-enhancing perspective, compartmentation, or the separation of personal information into small parts, is best achieved by the use of multiple biometrics.

Many public sector organizations use biometrics widely. For example, the U.S. Immigration and Naturalization Service (INS) currently makes the most extensive use of biometrics of all federal agencies.[2]

---

[1]See, e.g., Wayman, (1998). The use of multiple biometrics is sometimes referred to as "biometric diversity" or "biometric balkanization."

[2]See Appendix B, Program Reports. See also Wayman (2000). The INS Web site with public information regarding the use of one of its most popular biometric programs, known as the U.S. INS Passenger Accelerated Service System (INSPASS) is available at http://www.ins.usdoj.gov/graphics/howdoi/inspass.htm.

Other federal agencies using biometrics include the Federal Bureau of Investigation (FBI), the Federal Aviation Administration (FAA), and the Department of State and the Department of Defense (DoD). State social services programs use biometrics to reduce fraud and to enhance the convenience of these programs. Many state and local law enforcement agencies digitize fingerprints to speed criminal investigations and background checks. President Bill Clinton, in his 2000 State of the Union Address, and other political leaders have expressed interest in so-called "smart gun" technologies.[3] A smart gun could feature a biometric, such as a fingerprint, as an integral part of the firearm to make certain that only the authorized firearm user could fire the weapon. The private sector's growing interest in biometric applications stems as much from perceptions of increased efficiency and convenience as from increased security. In the Army, as in civilian life, biometrics are expected to be useful in many different applications.

Policymakers are also a driving force for those who view biometrics as a potential solution to the Army's access control problems. In 1999, Congress included $10 million in the Army's budget to study biometrics, specifically instructing the Army to conduct:

> [A]n immediate assessment of biometrics sensors and templates repository requirements and for combining and consolidating biometrics security technology and other information assurance technologies to accomplish a more focused and effective information assurance effort. (U.S. Senate, 2000)[4]

---

[3]See Clinton (2000). President Clinton stated:

> Technologies now exist that could lead to guns that can only be fired by the adults who own them. I ask Congress to fund research in smart gun technology. I also call on responsible leaders in the gun industry to work with us on smart guns and other steps to keep guns out of the wrong hands and keep our children safe.

See also LeDuc and Whitlock (2000) (reporting that Maryland Governor Parris Glendening is proposing to spend $3 million over the next three years to fund smart gun research), and Sinatra (2000).

[4]Public Law No: 106-79, Oct. 25, 1999. See also Byrd (1999), noting that "the Army has exhibited strong leadership in the exploration and development of technologies in the biometrics arena and is a natural leading candidate to be considered as the executive agent in this work for the Department of Defense and perhaps the federal government."

An important component of determining the feasibility of an Army biometrics program along the lines envisioned by Congress, is understanding what sociocultural (meaning sociological, legal, and ethical) concerns will be raised by Army use of biometrics and how the Army can best respond to these concerns. As with so many government-mandated programs, the use of biometrics requires trade-offs between individual rights and societal needs.

Sociocultural concerns are usually among the first to emerge during periods of change, particularly in response to an emerging technology, such as biometrics. Our societal code of ethics helps us adjust, or not, to the changes and challenges posed by a new technology. In the case of biometrics, many of these sociocultural concerns involve the appropriate protections of individual rights related to informational privacy, physical privacy, and religious beliefs. As the law mirrors the society that creates it, legal responses to such sociocultural concerns follow the sociocultural changes. These responses can include the enactment of new statutes or regulations, changes to existing ones, or the adoption of new codified restraints on behavior.

OBJECTIVES

As part of its response to the congressional directive to evaluate the feasibility of an Army biometrics program and center, Lieutenant General William H. Campbell, the Director for Information Systems, Command, Control, Communications, and Computers (DISC4) and the Army Chief Information Officer, asked RAND in October 1999 to review current commercially viable biometric applications and assess the sociological, legal, and ethical issues raised by Army use of biometrics, including establishment of a biometrics center.

APPROACH

To review commercially viable biometric applications, the RAND biometrics team consulted numerous paper and on-line publications and Web sites, interviewed more than 50 biometric experts in both the public and private sectors, and participated in several conferences at which a number of biometrics programs were presented. In sum, we reviewed approximately 50 biometric initiatives.

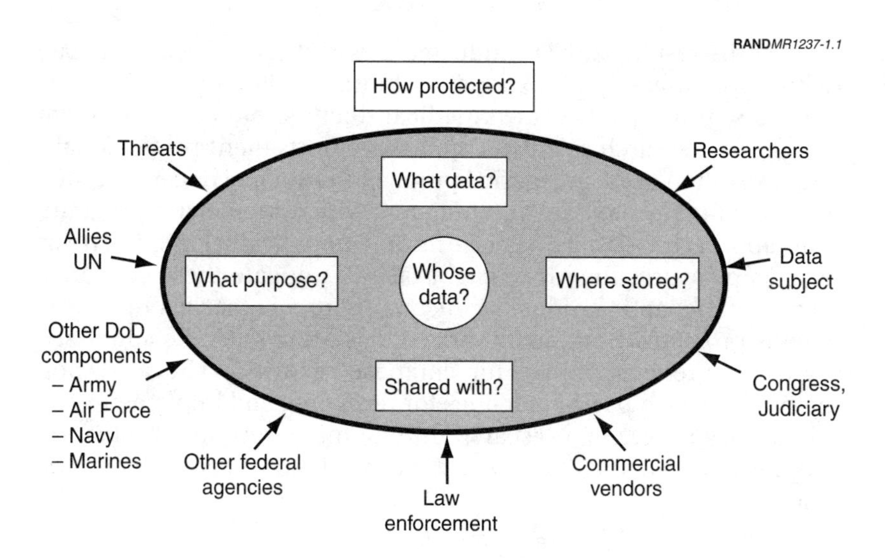

Figure 1.1—Dimensions of the Issues About Army Use of Biometrics

RAND assembled an interdisciplinary team to assess the sociological, legal, and ethical issues associated with biometrics and to develop a set of hypothetical issues that could stem from an Army biometrics program. We noted from the start the overlap among these areas and the important linkages between sociological, legal, and ethical issues. Figure 1.1 illustrates the dimensions of the issues raised and explored during our research.

To answer the many questions initially posed, we interviewed technologists, lawyers, ethicists, and privacy experts about individual rights and the interplay of law and technology. We also studied other biometric programs that would offer insights for how the Army community, consisting of servicemembers, Army civilians, contractors, Army retirees, and dependents, might respond to programs using biometrics.[5] In addition, we used information gleaned during our applications assessment to address some of our questions related to the capabilities of biometric technologies and, hence, the implications of the technologies for individuals and society.

---

[5]See Appendix B, which discusses a number of programs reviewed by RAND.

To test the results, RAND conducted a workshop in December 1999 with approximately 25 experts, including technologists, lawyers, ethicists, privacy advocates, medical doctors, biometric program managers, research scientists, and law enforcement professionals. The workshop used a scenario approach to explore concerns related to government-mandated compliance with a biometrics program. This approach focused on concerns that an individual might raise as well as measures the Army might take to mitigate these concerns. For example, an individual's concern about whether his biometric data is protected from unauthorized disclosure could be addressed by a comprehensive plan for database security. The workshop helped establish basic guidelines for designing and implementing a biometrics program, a research development, test, and evaluation (RDT&E) center, and, if necessary, a central repository for biometric data.

SCOPE

Although our approach was thorough, we recognize that our research had some limitations. First, our approach did not fully capture the views of every societal group. Second, our approach did not systematically survey the Army community, all of whom might be included in a biometric program. This issue is addressed again in Chapter Six where we discuss conclusions and recommendations. Third, our research focused primarily on the sociological, legal, and ethical issues of implementing an Army biometrics program in the United States. However, we are aware that foreign cultures and international legal systems will have different and perhaps contentious views of some of the issues we raise. While we briefly discuss European Union privacy laws in our legal assessment (see Appendix C), a systematic survey of the sociocultural concerns of individual nations and regional organizations is beyond the scope of this report. This issue is addressed further in our conclusions and recommendations.

ORGANIZATION OF THE REPORT

Following a brief overview of biometrics in Chapter Two, we use the next three chapters to answer three fundamental questions:

- What sociocultural concerns does the use of biometrics raise and how are these concerns different from those related to the use of other identification methods?
- What actions can the Army take to address these concerns?
- What is the feasibility of a national biometric center?

We finish the report with conclusions and recommendations that draw on our answers to these three questions.

We also include five appendices providing additional background material. Appendix A provides more detail on biometric technologies. Appendix B presents the experiences of some private and public sector biometric programs. Appendix C is a detailed review of the legal issues surrounding Army use of biometrics. Appendix D provides information on the U.S. government's Biometric Consortium, and Appendix E lists the names of the participants and organizations interviewed for the project.

Chapter Two

A PRIMER ON BIOMETRIC TECHNOLOGY

Here we provide a primer on biometric technology.[1] In particular, this chapter introduces the technical terminology and the major concepts related to biometric applications. We first define the terms, "biometrics" and "biometric authentication." Next, we discuss three universal components present in the operation of all biometric technologies, including the difference between applications that identify, or verify the claimed identity of, an individual. We then discuss some mainstream biometrics and their applications. We conclude with a table comparing salient characteristics of mainstream biometric technologies.

A DEFINITION OF BIOMETRICS AND BIOMETRIC AUTHENTICATION

A biometric is any *measurable*, *robust, distinctive, physical characteristic* or *personal trait* of an individual that can be used to *identify, or verify* the claimed identity of, that individual.

Measurable means that the characteristic or trait can be easily presented to a sensor and converted into a quantifiable, digital format.

---

[1]For those interested in a more detailed discussion about biometric technology, see Appendix A. This chapter does not cover standards for interoperability because this subject is tangential to RAND's project. In researching and writing this chapter, the authors relied heavily on the following sources: Hawkes and Hefferman (1999); Wayman (1999c); and Wayman (2000) (an updated version of "Testing and Evaluating Biometric Technologies: What the Customer Needs To Know" originally published in *Proceedings of the CardTech/SecurTech Conference '98*, May 1998). See also Dunn (1998).

This allows for the automated matching process to occur in a matter of seconds.

The *robustness* of a biometric is a measure of the extent to which the physical characteristic or personal trait is subject to significant changes over time. Such changes may occur because of the effects of an individual's exposure to chemicals, aging, or injury. A highly robust biometric is not subject to large changes over time, while a low degree of robustness indicates a biometric that could change considerably over time. For example, iris patterns, which change very little over a lifetime, are more robust than voices.

Distinctiveness is a measure of the variations or differences in the biometric pattern among the general population. The highest degree of distinctiveness implies a unique identifier, while a low degree of distinctiveness indicates a biometric pattern found frequently among the general population. The purpose of the biometric application determines the degree of robustness and distinctiveness required.

Identification differs significantly from *verification*. Identification is when the device asks and attempts to answer the question, "Who is X?" When biometrics are used to identify an individual, the biometric device reads a sample and compares that sample against every template in the database. This is called a "one-to-many" search (1:N). The device will either find a match and subsequently identify the person or not find a match and fail to identify the person.

Verification is when the device asks and attempts to answer the question, "Is this X?" after the user claims to be X. When biometrics are used to verify the claimed identity of an individual, the biometric device first requires input from the user. For example, the user claims his identity by using a password, token, or user name (or any combination of the three). The device also requires a biometric sample. It then compares the sample against the user-defined template (pointed to by the password, token, and/or user name) in the database. This is called a "one-to-one" search (1:1). The device will either find or not find a match between the two.

In general, there are three different approaches to recognizing an individual for security purposes, known as authentication. Presented in order of least secure and convenient to most secure and convenient, the first approach uses something you have, such as a token, card, or key. The second approach uses something you know, such as a password or PIN. The third uses something you are, a biometric. Any combination of these three further heightens security, while requiring all three provides the highest level of security.[2]

Biometric authentication refers to *automated methods of identifying or verifying* the identity of a *living person in real time* based on a *physical characteristic or personal trait.* The phrase, "living person in real time" is used to distinguish biometric authentication from forensics, which does not involve real-time identification of a living individual.

Biometric authentication technologies are used in two ways:

- To prove who you are or who you claim you are.
- To prove who you are not (for example, to resolve a case of mistaken identity).

KEY ELEMENTS OF ALL BIOMETRIC SYSTEMS

All biometric systems consist of three basic elements:

- enrollment,
- templates, and
- matching.

Enrollment is the process of collecting biometric samples from a person and the subsequent generation of a template. Typically, the device takes three samples of the same biometric and then averages them to produce an enrollment template. Templates are the data representing the enrollee's biometric. They are created by the biometric device, which uses a proprietary algorithm to extract "features" appropriate to that technology from the enrollee's sam-

---

[2]Security also depends on such factors as the care taken to apply security measures properly, insofar as safeguarding tokens and passwords and ensuring that transmissions of biometric data are adequately protected.

ples.[3] These features are also referred to as minutiae points for some technologies, such as fingerprint systems. Because templates are only a record of distinguishing features of a person's biometric characteristic or trait (and not an image or complete record of the actual fingerprint or voice), the template is usually small and allows for the near-instantaneous processing time characteristic of biometric authentication. The small size of some templates allows for storage on magnetic stripes or bar codes placed on plastic cards or smart cards.[4] An example of the formation of a template for a fingerprint is shown in Figure 2.1.

For any biometric technology, a small percentage of the population will be unable to produce a usable template. This failure to enroll (or acquire) is the failure of the technology to extract adequate distinguishing features appropriate to that technology. For example, a small fraction of the population cannot be fingerprinted either because their prints are not distinctive enough (e.g., no bifurcations that can be picked up by the system) or because of the individual's occupation or age, which can alter distinguishing features.

Matching is the process of comparing a submitted biometric sample against one (verification) or many (identification) templates in the system's database.

In general, verification applications provide more security than identification applications because a biometric and at least one other piece of input (e.g., PIN, password, token, user name) are required to match a template. Verification provides a user with control over his own data and over the biometric authentication process, provided

---

[3]Image files of fingerprints also may be of interest to the Army because of their law enforcement, forensic, and technical applications. In the case of fingerprints, the Army may want to keep both electronic image files of the fingerprints as well as the biometric templates. While the image files are too large to be used for biometric applications, they would be useful for criminal investigation and other forensic purposes. Moreover, the Army might want to store image files to provide greater technical flexibility. For example, if the Army did not keep image files of enrollees, it might have to physically reenroll each individual if the Army decided to change to a different proprietary biometric system. Any image file is also referred to as raw data or the *corpus*.

[4]Appendix A includes a detailed discussion about template storage.

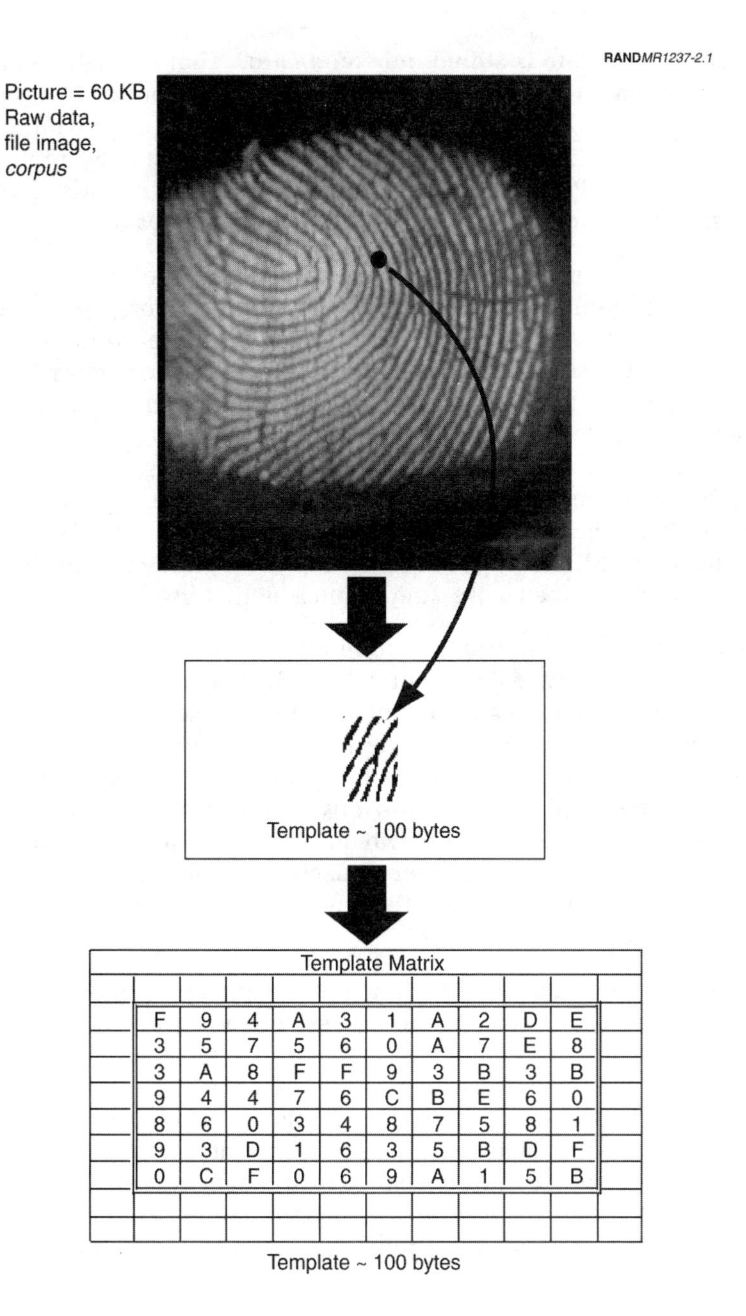

Figure 2.1—Example of the Formation of a Template for a Fingerprint

that the template is stored only on a card. That is, such a system would not allow for clandestine, or involuntary, capture of biometric data because the individual would know if he were providing the card. Because the search seeks only a match against one template in the database, verification applications require less processing time, less memory, and less cost than identification applications.

Accuracy and error rates must be examined by the end-user when choosing biometric devices. When discussing errors, we prefer to use the terms, false match rate (FMR), and false nonmatch rate (FNMR). A false match occurs when a sample is incorrectly matched to a template in the database. A false nonmatch occurs when a sample is incorrectly not matched to a truly matching template in the database (i.e., a legitimate match is denied). These two terms are often misnamed "false acceptance rate" and "false rejection rate," respectively, but these terms are application-dependent in meaning. FMR and FNMR are application-neutral terms to describe the matching process between a live sample and a biometric template.

Identification applications require a highly robust and distinctive biometric, otherwise the error rates falsely matching and nonmatching users' samples against templates breaches security and inhibits convenience. Applications where the end-user wants to identify criminals (immigration, law enforcement, etc.) or other types of "wolves in sheep's clothing" must use an identification application. Other types of applications may require a verification application. In many ways, deciding whether to use verification or identification requires a balance between the end-user's needs for security and convenience.

Template management is an integral component of balancing privacy, security, and convenience issues. All biometric systems face a common issue: The template database must be stored somewhere. Biometric templates must be protected to prevent identity fraud and maintain user privacy. Possible solutions include storage on the biometric device itself, a central computer that is remotely accessed, a plastic card or token with a bar code or magnetic stripe, Radio Frequency Identification Device (RFID) cards and tags, optical memory cards, PCMCIA (Personal Computer Memory Card International Association) cards, and smart cards.

Smart cards are the size of credit cards and have a microchip or microprocessor chip embedded in them. The chip can store a variety of electronic data, including a biometric template that can be protected using biometric authentication. Smart cards come in two types: contact and contactless smart cards. A contact smart card must be inserted into a smart card reader to be used. A contactless smart card need only be placed near an antenna to carry out a transaction.

An important security issue with regard to template database management is whether the database will serve a unique purpose or if it will be used for multiple purposes. For example, a facilities manager might use a fingerprint reader to control building access. He might also want to use the same fingerprint template database to identify employees logging onto their computer network. The manager should consider several important questions, such as should separate databases be used for these different purposes and is it an acceptable risk to access employee fingerprints from a remote location for multiple purposes? The transmission of data across wires to a central database presents risks that the biometric template might be captured or stolen.

An additional privacy and security concern is what additional personal information will be stored about each user with his biometric template and whether his biometric is used to link to other personal information about him.

MAINSTREAM BIOMETRICS AND THEIR APPLICATIONS

Of the many possible biometrics, at least eight mainstream biometric authentication technologies have been deployed or pilot-tested in applications in the public and private sectors. These are fingerprint, hand/finger geometry, facial recognition, voice recognition, iris scan, retinal scan, dynamic signature verification, and keystroke dynamics.[5] Each is discussed briefly below.

---

[5]For a comprehensive examination of mainstream biometrics, see Jain, Bolle, and Pankanti (1998).

Fingerprint

The fingerprint biometric is an automated digital version of the old ink-and-paper method used for more than a century for identification, primarily by law enforcement agencies. The biometric device involves a user placing his finger on a platen for the fingerprint to be read. The minutiae are then extracted by the vendor's particular algorithm to create a template. Fingerprint biometrics have three main application arenas: large-scale Automated Finger Imaging Systems (AFIS) for law enforcement uses, fraud prevention in entitlement programs, and access control for facilities or computers.

Hand/Finger Geometry

Hand or finger geometry is an automated measurement of many dimensions of the hand and fingers. Neither of these methods take prints of the palm or fingers. Rather, only the spatial geometry is examined as the user lays his hand on the sensor's surface and uses guiding poles between the fingers to place the hand properly and initiate the reading. Finger geometry typically uses two or three fingers. During the 1996 Summer Olympics, hand geometry secured access to the athletes' dorms at Georgia Tech. Hand geometry is a well-developed technology that has been thoroughly field-tested and is easily accepted by users.

Facial Recognition

Facial recognition is an automated method to record the spatial geometry of distinguishing features of the face. Different methods of facial recognition among various vendors all focus on measures of key features. Noncooperative behavior by the user and environmental factors, such as lighting conditions, can degrade performance for facial recognition technologies. Facial recognition has been used in projects designed to identify card counters in casinos, shoplifters in stores, criminals in targeted urban areas, and terrorists overseas.

Voice Recognition

Voice recognition is an automated method of using vocal characteristics to identify individuals using a pass-phrase. The technology itself is not well-developed, partly because background noise affects its performance. Additionally, it is unclear whether the technologies actually recognize the voice or just the pronunciation of the pass-phrase (password) used to identify the user. The telecommunications industry and the National Security Agency (NSA) continue to work to improve voice recognition reliability. A telephone or microphone can serve as a sensor, which makes this a relatively cheap and easily deployable technology.

Iris Scan

Iris scanning measures the iris pattern in the colored part of the eye (although the color has nothing to do with the scan). Iris patterns are formed randomly. This means no two iris patterns are the same; the iris pattern of one's left eye is different from the iris pattern of the right eye. Iris scans can be used for both identification and verification applications. ATMs ("Eye-TMs"), grocery stores (for checking-out), and the Charlotte/Douglas International Airport (physical access) use iris scanning in test applications. During the 1998 Winter Olympic Games in Nagano, Japan, an iris scanning identification system controlled access to the rifles used in the biathlon.

Retinal Scan

Retinal scans measure the blood vessel patterns in the back of the eye. The device involves a light source shined into the eye of a user who must stand very still within inches of the device. Because the retina can change with certain medical conditions, such as pregnancy, high blood pressure, and AIDS, this biometric has the potential to reveal more about individuals than only their identity. Because users perceive the technology to be intrusive, retinal scanning has lost popularity with end-users.

Dynamic Signature Verification

Dynamic signature verification is an automated method of examining an individual's signature. It uses a stylus and surface on which a person writes. This technology examines dynamics, such as speed, direction, and pressure of writing; time that the stylus is in and out of contact with the "paper"; total time of the signature; and where the stylus is raised and lowered onto the "paper."

Keystroke Dynamics

Keystroke dynamics is an automated method of examining an individual's keystrokes on a keyboard. The technology uses a keyboard compatible with PCs. This technology examines such dynamics as speed and pressure, total time of typing a particular password, and the time a user takes between hitting certain keys. Keystroke dynamics has the potential for continuous authentication of identity while a person is using a computer.

SALIENT CHARACTERISTICS OF MAINSTREAM BIOMETRICS

Table 2.1 compares the eight mainstream biometrics in terms of several characteristics, ranging from how robust and distinctive they are to what they can be used for (i.e., identification or verification or verification alone).[6]

When we compare how mainstream biometrics can be used, we find that about half can be used reliably for either identification or verification purposes. The other half are best only for verification purposes. In particular, hand geometry has only been used for verification applications, such as physical access control and time and attendance. Biometrics that can only be used for verification purposes present fewer privacy concerns because they are not trackable.

The robustness and distinctiveness of biometrics vary considerably. Fingerprinting is moderately robust, and, although it is distinctive, a small percentage of the general population at any given time has unusable prints. While hand/finger geometry is moderate on the

---

[6]This table is an effort to assist the reader in categorizing biometrics along important dimensions. Because this industry is still establishing standards and the technology is changing rapidly, it is difficult to make unequivocal assessments. The table represents our assessment based on discussions with technologists, vendors, and program managers. This table is compiled from various sources, including Jain, Bolle, and Pankanti (1998) and various presentations made at the SJB Biometrics 99 Workshop, November 9-11, 1999, particularly Hawkes and Hefferman (1999).

robustness scale, it is not very distinctive. Facial recognition is neither highly robust nor distinctive. As for voice recognition, assuming the voice and not the pronunciation is what is being measured, this biometric is moderately robust and distinctive. Iris scans are both highly robust, because they are not susceptible to day-to-day changes or damages, and distinctive, because they are randomly formed. Retinal scans are also fairly robust and very distinctive. Finally, neither dynamic signature verification nor keystroke dynamics are particularly robust or distinctive.

As the table shows, the biometrics vary in terms of how intrusive they are, ranging from those biometrics that require touching to others that can recognize an individual from a distance.

Table 2.1

Comparison of Mainstream Biometrics

Biometric	Identify versus Verify	How Robust	How Distinctive	How Intrusive
Fingerprint	Either	Moderate	High	Touching
Hand/Finger Geometry	Verify	Moderate	Low	Touching
Facial Recognition	Either	Moderate	Moderate	12+ inches
Voice Recognition	Verify	Moderate	Low	Remote
Iris Scan	Either	High	High	12+ inches
Retinal Scan	Either	High	High	1–2 inches
Dynamic Signature Verification	Verify	Low	Moderate	Touching
Keystroke Dynamics	Verify	Low	Low	Touching

Chapter Three

WHAT CONCERNS DO BIOMETRICS RAISE AND HOW DO THEY DIFFER FROM CONCERNS ABOUT OTHER IDENTIFICATION METHODS?

Any biometrics program must be prepared to deal with individuals who cannot or will not participate in the program. Some people, through no fault of their own, cannot provide the chosen biometric because they have unmeasurable fingerprints or eyes, for example. Thus, all biometric systems have a small number of people who simply cannot be enrolled.

Others, however, deliberately choose not to participate in biometric programs because of their individual beliefs. While these individuals constitute a relatively small minority, they are the most likely contingent to voice their concerns. Their criticism will probably not cause a biometric program to collapse or render it ineffective. However, such criticism could inhibit a biometric program's development, implementation, and support. Thus, the Army must bear in mind that biometric technology is not without its critics,[1] and it must take into account current and potential sociological concerns.

In this chapter, we answer the question of what sociocultural concerns biometrics raise and whether these concerns differ from those related to other identification methods. In response to the first part of the question, we identify and briefly discuss three key sociocultural concerns: informational privacy, physical privacy, and reli-

---

[1]See, e.g., Garfinkel (2000), discussing privacy concerns of new technologies including biometrics, and Woodward (1997a), surveying the privacy enhancing and privacy threatening aspects of biometrics.

gious objections. With regard to the second part of the question, we find that while similarities exist among the concerns, differences exist between biometrics and other identification methods.[2]

Background. In the United States, the freedom of the individual is perceived to be closely related to his ability to operate somewhat autonomously and anonymously in the eyes of the state as well as other organizations. For example, we cast secret ballots; we have certain legal rights to decisional, informational, and physical privacy; and we have placed constraints on the sharing of personal information about us within the federal government. Although constraints on the sharing of information about us by other organizations are more limited, we, as a society, are grappling with how best to respond to the capabilities information technology affords.

Biometric applications have the potential to further reduce anonymity. In addition, as with any technological innovation, some people will find certain aspects of biometrics discomforting or unacceptable for a variety of sociocultural reasons. Some of these reasons may be recognized by our societal ethics and laws, such as those protecting religious freedom and privacy rights. However, as a society, we acknowledge, and the laws reflect, an expectation that in some cases the needs of society will override individual objections to participating in an Army biometrics program or any other government-mandated biometrics program. If the discomfort with such a program seems to arise from unfamiliarity with a new technology, as opposed to deep-seated moral or religious objections, the decision about whether to force compliance with the new program must weigh the importance of the societal need for the new program against societal concerns for individual rights. In a type of utilitarian calculus, the law also recognizes the necessity to balance the needs of the whole against the rights of the individual.

A review of past and current biometric programs suggests that the use of biometrics in the United States evokes several sociocultural concerns. These concerns may be based on a variety of factors, including fears about the centralization of biometric identification information and the potential for misuse of these data, concerns

---

[2]Biometric programs provided important insights to this part of our research. Descriptions of some may be found in Appendix B.

about the physical intrusiveness of the technology, and religious objections to the technology's use.[3]

These issues will more fully evolve over time and may change significantly as biometrics are introduced more broadly through private sector and public sector applications. It seems likely that as biometrics become more pervasive, the research community will begin to determine whether these measures can reveal more about a person than only his identity. For example, knowing that certain medical disorders are associated with specific fingerprint abnormalities, researchers might actively investigate such questions as, can fingerprint template patterns be linked to behavioral characteristics, or predispositions to medical conditions? If these questions are answered affirmatively, biometrics might become not only an identifier, but also a source of information about an individual. Such a development would likely have a significant impact on how biometrics are perceived and managed in the United States and abroad.

KEY SOCIOCULTURAL CONCERNS

The key sociocultural concerns associated with biometrics fall into three main categories:

- Informational privacy.
- Physical privacy.
- Religious objections.

Informational Privacy

The most significant informational privacy concerns relate to the threat of function creep and the tracking capabilities of biometrics. These concerns are addressed below.

Function Creep. Function creep, or mission creep, is the process by which the original purpose for obtaining the information is widened

---

[3]See Appendix B, Program Reports, Citicorp Clip Card and Security Infrastructure and Columbia Presbyterian Hospital, for examples of privacy concerns raised during program implementation.

to include purposes other than the one originally stated. Function creep can occur with or without the knowledge or agreement of the person providing the data. Many privacy experts contend that function creep is inevitable.[4]

Depending on whom it affects and how it affects them, function creep may be seen as desirable or undesirable. For instance, using SSNs to search for a parent who is delinquent with child support payments may be seen as desirable. On the other hand, having a person's digitized state Department of Motor Vehicles (DMV) photograph sold to a commercial firm to create a national photo ID database might be considered unacceptable (Davies, 1994, pp. 61–62).[5]

Additional purposes can be useful and valuable to society, but ethical concerns arise when biometrics are used beyond their original purpose, without the informed and voluntary consent of the participants. These concerns include whether participants have the right to reassess their participation given the new purpose for the data, the implications of a decision not to cooperate in providing biometric data, and justification of the new purposes, given the program's original intent.

Tracking. The use of massive databases containing detailed personal information in both the public and private sectors has raised concerns about an individual's ability to maintain his or her anonymity (Garfinkel, 2000, pp. 1–36). Some people fear a "Big Brother" government able to track every individual. Tracking, which may be thought of as a particular type of function creep, refers to the ability to monitor in real time an individual's actions or to search databases

---

[4]For example, Simon G. Davies (1994), the Director of Privacy International, has explained:

> The history of identification systems throughout the world provides evidence of "function creep"—application to additional purposes not announced, or perhaps even intended, at the commencement of the scheme. . . . The existence of a relatively high-integrity scheme would create irresistible temptations to apply it widely, and interrelate many hitherto separate collections of personal information.

[5]In fact, South Carolina sold photographs of the state's drivers to Image Data LLC, a New Hampshire company.

that contain information about these actions. For example, if an individual must use a standard biometric for multiple governmental, business, and leisure transactions of everyday life, it becomes possible that each of these records could be linked through the standardized biometric. This link could allow an entity, such as the government, to compile a comprehensive profile of the individual's actions. This Big Brother concern has been expressed by privacy expert Roger Clarke (1994, p. 34):

> Any high-integrity identifier [like biometrics] represents a threat to civil liberties, because it represents the basis for a ubiquitous identification scheme, and such a scheme provides enormous power over the populace. All human behavior would become transparent to the state, and the scope for non-conformism and dissent would be muted to the point envisaged by the anti-Utopian novelists.

The possibility of clandestine capture of biometric data increases concerns about Big Brother. For example, facial recognition systems can track individuals without the individual's knowledge or permission. This alone raises ethical concerns. Moreover, the information from tracking can be combined with other personal data, acquired through biometrics or other means, to provide even more insight into an individual's private life.[6]

Misuse of Data. Misuse of personal information, including the stealing of identities (identity theft), has become more of a threat as information technology, including electronic commerce, has become ubiquitous. Used in certain ways, biometrics provide greater security because the biometric identifier is much harder to steal or counterfeit. As sociologist Amitai Etzioni (1999, p. 125, emphasis added) has explained:

> Reliable identifiers could replace the existing patchwork of passwords that are often forgotten, lost, or misappropriated. The same identifiers could be used to ensure that one's vote is not forged, that one's credit card is not misused, that one's checks are not cashed by

---

[6]Earlier this year, controversy surrounded the disclosure that law enforcement used facial recognition to surreptitiously surveil spectators at the Super Bowl in Tampa, Florida, in an effort to identify would-be criminals and terrorists. See, e.g., Woodward (2001), Piller et al. (2001), and Slevin (2001).

> others. . . . In short, reliable universal identifiers—*especially biometric ones*—could go a long way toward ensuring that people are secure in their identity, thereby allowing others to trust that they are who they claim to be.

On the other hand, where biometrics are authenticated remotely, that is, by transmission of data from a sensor to a centralized data repository, a hacker might be able to steal, copy, or reverse-engineer the biometric.[7] This misappropriation could also come about through insider misuse—e.g., the rogue employee. Without proper safeguards, files could be misappropriated and transactions could be performed using other people's identities.

Physical Privacy

The use of biometrics may raise physical privacy concerns. These concerns are threefold: the stigma associated with some biometrics, such as fingerprints; the possibility of actual harm to the participants by the technology itself; and the concern that the devices used to obtain or "read" the biometric may be unhygienic.

Stigmatization. Concerns about stigma vary tremendously in society. In the United States, some individuals and segments of society associate fingerprinting with law enforcement, acts of criminal behavior, and oppressive government (Garfinkel, 2000, pp. 43–44). However, among the voluntary private sector programs we reviewed that used fingerprints (see Appendix B), no program managers cited this stigma as a concern among participants. Stigmatization may be more of an issue for participants in mandatory programs, such as those the Army would implement, as well as biometric applications that have been implemented by state departments of social services. The program managers we spoke with, however, indicated that these objections were easily overcome through education about the protections that would be in place on using and safeguarding biometric data.[8]

---

[7]See Appendix B, Program Reports, Fort Sill Program, as an example of concerns over reverse-engineering fingerprints.

[8]The outreach and public education programs of the state social services departments generally focused on the benefits for social services clients by having a biometric

All service members and applicants for federal employment must provide fingerprints to the FBI as part of a background check. Hence, to this group, fingerprinting is nothing new. Based on our discussions with international privacy experts and program managers, it appears that fingerprinting is more of an issue in other nations and cultures, although several foreign biometric programs use fingerprints and report little concern about social stigma among their populations.

Actual Harm. Concerns about actual harm that could be caused by biometric technologies are primarily perceptual. Although the technology is in fact harmless, the perception of harm may cause users to obstruct the implementation of a program or be reluctant to participate in it. For example, we can imagine military pilots, whose careers depend on their eyesight, being greatly concerned about a biometrics program that requires them to look in close proximity at a device to have their retina scanned. Others may be concerned that a dismembered limb could be stolen and used to "fool" a system.

Hygiene. Objections to biometrics based on concerns about the cleanliness of sensors is another issue. Much as with concerns of the cleanliness of public restrooms, participants may feel uncomfortable placing their faces against a machine to have their retinas scanned after many others have done so or touching a hand-geometry scanner during flu and cold season. However, we know of no biometric application overturned for hygienic reasons.

The degree to which objections based on physical requirements might arise, if at all, are likely to be correlated with the biometric technology chosen, the size of the group using the biometric, and whether the sensor is shared by many (as with a hand-geometry reader at an airport) or used individually (as with a desktop computer fingerprint sensor).

---

identifier as an alternative to requiring the clients to produce numerous paper documents. In addition, the education programs emphasized the departments' commitments to keep biometric information from law enforcement officials. Officials found that as clients began using biometrics they realized that it made the process easier. Clients also felt better because they were not always having to prove their identity—often difficult for the poor without drivers' licenses, credit cards, and other standard forms of identification. See Appendix B, Program Reports, Social Services.

Religious Objections

In the United States, religious objections to biometrics might arise from a variety of different groups.[9] For example, certain Christians interpret biometrics to be a "Mark of the Beast." The objection is based on language in "Revelation":

> [The Beast] causeth all, both small and great, rich and poor, free and bond, to receive a mark in their right hand, or in their foreheads: And that no man might buy or sell, save that he had the mark, or the name of the beast, or the number of his name. . . . and his number is six hundred, threescore, and six. (Revelation, 13:16–18.)

Certain Christians consider the biometric to be the brand discussed in Revelation and biometric readers as the only means of viewing these brands. Similarly, M. G. "Pat" Robertson, host of "The 700 Club" and founder of The Christian Broadcasting Network, Inc., observes that the "Bible says the time is going to come when you cannot buy or sell except when a mark is placed on your hand or forehead." He expresses doubts about biometrics and notes how the technology is proceeding according to Scripture (700 Club, 1995).

Religious objections have arisen when identification programs have been implemented. In Alabama, two people objected to providing an SSN, as required under Alabama law, to apply for a driver's license. The individuals based their refusal on their sincerely held religious beliefs that prevent them from having an SSN. This case is pending in the Alabama state courts (*Alabama Lawsuit*, 2000). In the case of biometric identification, religious objections contributed, at least in part, to the failure of Alabama DMV's fingerprint program. In Alabama, groups including the Christian Coalition, Southern Christian Leadership Conference, and the American Civil Liberties Union, vigorously protested an effort to place a fingerprint biometric on all driver's licenses. In July 1997, Alabama Governor Fob James, Jr., stopped the proposed program because of these objections (Stamper, 1997). However, five other states—California, Texas, Colorado, Hawaii, and Georgia—successfully require a driver to provide a fingerprint on the driver's license, without significant public oppo-

---

[9]See Appendix B, Program Reports, Connecticut Department of Social Services and Fort Sill, Oklahoma.

sition. In West Virginia and the District of Columbia, providing a fingerprint for the license is optional. Moreover, another dozen or so states are considering or planning to use fingerprints on driver's licenses in the near future (Woodward and Smythe, 2000).

We do not expect religious objections to biometrics to be widespread, but such objections must be taken seriously because of societal and legal emphasis on respect for sincerely held religious beliefs.

BIOMETRICS RAISE SIMILAR YET DIFFERENT CONCERNS

When considering whether sociocultural concerns about biometrics are similar to or different from those raised about the use of other identifiers, we find that many of the concerns are very similar.

Informational privacy concerns are not new. The use of SSNs is a prime example of function creep—an individual's SSN is used for an array of purposes in the public and private sectors. Tracking issues are also a major concern today in part because the public has been made aware of informational commerce or the profitable use of data as a commodity.[10] For example, many retailers currently sell personal information collected from customers to data merchants, who compile and sort data from multiple sources into more comprehensive and marketable databases. At issue is not the loss of your grocery store account number but that all information associated with it, namely your food, beverage, and other purchasing habits, are now known by many unidentified people or organizations.

Religious objections are not unique to biometrics technology. Certain individuals have opposed SSNs on religious grounds.[11] In addition, the Army has had to address a variety of other religious accommodation issues, ranging from uniform attire to religious practice.

---

[10]See, e.g., *Economist* (1999). See also O'Harrow, (1999b), noting that Acxiom Corp., an Arkansas company that provides information to marketers, has amassed 135 million consumer telephone numbers, including about 20 million that are unlisted, to help identify and profile people who call toll-free lines to shop or make an inquiry.

[11]See *Bowen v. Roy*, 476 U.S. 693 (1986) (Parents contended that obtaining a SSN for their daughter would violate their Native American religious beliefs). See also *Alabama Lawsuit* (2000).

However, despite these similarities, biometric technologies raise concerns different from traditional identifiers. The concerns are a product of differences in the biometrics technology itself, the data produced by the technology, and how such data might be used. Theft of a biometric identifier poses a new set of issues.[12] If an individual's PIN for his ATM is stolen, the bank simply issues the individual another one and cancels transactions under his name made using the old PIN. However, if an individual's biometric is stolen, there must be a system in place to accept an alternative biometric or means of identification.

In addition, the data that some biometric technologies produce are, or have the potential to be, different from what is produced by traditional identifiers. Some biometric information contains medical information.[13] With technological advancements, medical information may someday be available using biometrics. Because biometrics are inherent to the individual, researchers are likely to try to link medical predispositions, behavioral types, or other characteristics to particular biometric patterns. This possibility makes biometrics different from PINs, passwords, and other generated numbers used to identify an individual.

Finally, biometrics present a greater potential for function creep because biometrics offer an ability to track individuals in a way that current passwords and PINs cannot. When biometrics replace or enhance existing security systems, their role is no different from current techniques that identify an individual or verify that a person is who he or she claims to be.[14] The difference is that unlike other

---

[12]See, e.g., Garfinkel (2000, pp. 62–65).

[13]Recent scientific research suggests that fingerprints disclose medical information. Chen (1998, pp. 221–226) states, "Certain chromosomal disorders are known to be associated with characteristic dermatoglyphic abnormalities." He specifically cites Down's syndrome, Turner's syndrome, and Klinefelter's syndrome as chromosomal disorders that cause unusual fingerprint patterns in a person. DNA is an example of a biometric that contains much more than simple identification information. However, as of 2001, DNA analysis is not sufficiently automated or quick enough to be viable for use in a biometric program.

[14]For example, several state social services agencies use biometrics to verify the identity of entitlement recipients as part of their fraud prevention programs. The problem is that some people illegally establish multiple identities and collect multiple entitlement payments, known as "double-dipping." While procuring fake documentation sufficient to establish an identity is not difficult, use of a biometric identifier

forms of identification, which are specific to particular purposes (generally a transaction of some sort), all individuals provide their biometrics as they go about their daily tasks. Faces, fingerprints, and voices are available for all to see and recognize. As a result, biometric technologies could make it feasible to capture this information and track people without their knowledge or consent. The use of biometrics as an identifier further magnifies this concern, because the biometric is not something that can be changed or discarded. For the reasons discussed above, biometrics may be perceived by some as a qualitatively different means of checking identity. Because the technology is new, however, perceptions are likely to change over time.

---

would reveal that the multiple identities all use the same biometric: a clear sign of fraud. Similarly, if a credit card were protected by a biometric, a thief could still steal it from the authorized user, but unless the thief could produce the biometrics of the authorized user for the sensor at the point of sale, the card would not be accepted.

Chapter Four

WHAT STEPS CAN THE ARMY TAKE TO ADDRESS THESE CONCERNS?

Given the three major sociocultural concerns associated with the Army's use of biometrics, namely protecting informational privacy, safeguarding physical privacy, and addressing religious objections, this chapter focuses on how the Army can most effectively address these concerns. Because biometrics are similar in many ways to more traditional identifiers, such as photographs, the Army can look to existing laws, regulations, and precedents to address the main sociocultural concerns. This body of law suggests that the Army has a solid framework in place to address sociocultural concerns.

However, in light of the novelty of the technology and the heightened interest among citizens in informational privacy, it may be prudent for the Army to address concerns about biometrics within the context of a broader approach. By doing so, the Army can reassure its community and the American public that it takes the use of biometrics and protection of privacy and religious freedoms seriously, its biometrics program is well-thought-out, it has taken reasonable precautions to protect personal data, and it has made choices about the technologies being used and the structure of the program with consideration for their effect on individuals.

In this chapter, we begin by discussing legal precedents and related policies and procedures the Army might apply to address the specific sociocultural concerns discussed in Chapter Three.[1] We conclude by

[1] Readers interested in a more detailed legal assessment should refer to Appendix C.

discussing what a comprehensive approach to managing these concerns ought to entail.

PRIVACY ACT OF 1974: A BASELINE FOR ADDRESSING SOME SOCIOCULTURAL CONCERNS

Personal information in the hands of the Army is not a new issue. Accordingly, many of the sociocultural concerns raised by the use of biometrics can be addressed through Army policies and procedures established to carry out the requirements of the Privacy Act of 1974. The Privacy Act regulates the collection, maintenance, use, and dissemination of personal information by federal government agencies, including DoD and the Army. Although the Act does not specifically mention biometrics, analysis suggests that many Army biometric programs would fall under this Act, in particular those biometric applications involving "a system of records."[2]

The Privacy Act gives certain rights to the individual who provides personal information and places certain responsibilities on the agency collecting the personal information. While the Privacy Act addresses informational privacy concerns, it does not address physical privacy and religious freedom concerns.

The Privacy Act's basic provisions, reflected in both the DoD Privacy Program and the Army Privacy Program,[3] include the following:

- Restricting federal agencies from disclosure of personally identifiable records maintained by the agencies.
- Requiring federal agencies to maintain records with accuracy and diligence.

---

[2]The Privacy Act applies to a "record" that is "contained in a system of records." While the Act's definition of record includes any "other identifying particular assigned to the individual such as a finger or voice print or a photograph" (see 5 U.S.C. § 552a(a)(4)), not all biometric programs are necessarily contained in a system of records. For example, the Fort Sill test of a biometrically protected smart card (see Appendix B) was not contained in a system of records because the biometric was stored only on the card and not in any central database. The Fort Sill test is an example of a biometric being used to enhance privacy.

[3]Unless otherwise indicated, the Privacy Act provisions discussed here apply to DoD and the Army.

- Granting individuals increased rights (1) to gain access to records maintained on them by federal agencies and (2) to amend their records provided they show that the records are not accurate, relevant, timely, or complete.
- Requiring federal agencies to establish administrative, technical, and policy safeguards to protect security of records.[4]

To ensure compliance with the Act, DoD and the Army have established several institutional assets to help oversee policies and procedures related to privacy. For example, DoD has a Defense Privacy Board and supporting staff in the Defense Privacy Office.[5]

The Act requires those agencies establishing systems of records to publish in the *Federal Register* information about the systems of records in their charge, give individual notice of the uses to which the data will be put, and safeguard data. The Army now has 249 systems of records for which notice must be published. These systems range from "Official Personnel Folders and General Personnel Files" to "Individual Health" to "Carpool Information/Registration System."[6] As Army use of biometrics will likely lead to the establishment of new systems of records and revisions to old systems, the Army must comply with this Privacy Act Systems of Records Notice requirement.

The Act's requirement that individuals be given notice addresses many of the concerns raised in Chapter Three. The notice must state the authority sanctioning the solicitation of the information, the purpose for which the information is intended, the routine uses that may be made of the information, whether the data collection effort is voluntary or mandatory, and the implications to the data subject of failing to provide the requested information.[7]

---

[4]See, e.g., Cate (1997, p. 77).

[5]The Army can draw on the many existing institutional assets who have extensive experience in Privacy Act matters. These assets include the Defense Privacy Board, the Assistant Secretary of Defense (Comptroller), the Defense Privacy Office, the DoD General Counsel, the Army Assistant Chief of Staff for Information Management, the Army General Counsel, the Army Judge Advocate General, the Deputy Chief of Staff for Personnel (DCSPERS), OMB, and many others.

[6]See Department of the Army (2000).

[7]See 5 U.S.C. § 552a(e)(3)(A)-(D). The authority may be granted by statute or executive order of the President. See 5 U.S.C. § 552a(e)(3)(A).

Although the Act prohibits a federal agency from "disclos[ing] any record without the consent of the individual to whom the record pertains," it provides for certain disclosure exceptions listed below.[8] These twelve exceptions to the "No Disclosure Without Consent" Rule are:

- the "Intra-Agency Need to Know" Exception,
- the "Required Freedom of Information Act (FOIA) Disclosure" Exception,
- the "Routine Use" Exception,
- the "Bureau of the Census" Exception,
- the "Statistical Research" Exception,
- the "National Archives" Exception,
- the "Law Enforcement Request" Exception,
- the "Individual Health or Safety" Exception,
- the "Congressional" Exception,
- the "General Accounting Office" Exception,
- the "Judicial" Exception, and,
- the "Debt Collection Act" Exception.

Some privacy advocates contend that the routine use exception has been used by federal agencies to justify almost any use of the data (Cate, 1997, p. 78, footnote omitted). For example, the Army has a so-called "law enforcement blanket routine use," which applies to every record system maintained by the Army, unless specifically stated otherwise. The law enforcement blanket routine use allows the Army to share routinely any record indicating a potential violation of the law with the appropriate federal, state, local, or foreign agency charged with investigating the matter.[9] Similarly, the Army

[8]See 5 U.S.C. § 552a(b).

[9]See, e.g., OMB Guidelines, 40 Fed. Reg. at 28,955 (proper routine use is "transfer by a law enforcement agency of protective intelligence information to the Secret Service"). See also 28 U.S.C. § 534 (authorizing Attorney General to exchange criminal records

routinely discloses any records indicating a possible violation of law, regardless of the purpose for collection, to law enforcement agencies for purposes of investigation and prosecution.[10] On the other hand, the Army can exempt certain programs from this "law enforcement blanket routine use." For example, it does not apply to the "Armed Forces Repository of Specimen Samples for the Identification of Remains" System of Records, which includes "specimen collections from which a DNA typing can be obtained."[11]

The Army does not necessarily have the final say over how its data will be used or shared. Congress can always mandate additional new "routine uses" for data. For example, Congress has mandated the establishment of a federal "Parent Locator Service" and requires federal agencies to comply with requests from the Secretary of Health and Human Services for addresses and places of employment of absent parents.[12]

As noted above, the Privacy Act also requires the federal agency to put in place appropriate safeguards to protect information in its databases.[13] Both the agency and the employee responsible for any breach can be found legally liable for a Privacy Act violation, including civil liability for the agency and criminal liability for the individual.

In *Whalen v. Roe*, the Supreme Court directly addressed informational privacy concerns.[14] *Whalen* involved the constitutional question of whether the State of New York could record and store, in

---

with "authorized officials of the Federal Government, the States, cities, and penal and other institutions").

[10]See OMB Guidelines, 40 Fed. Reg. at 28,953; see also 28 U.S.C. § 535(b) (1994) (requiring agencies of the executive branch to expeditiously report "[a]ny information, allegation, or complaint" relating to crimes involving government officers and employees to the U.S. Attorney General).

[11]See 63 Fed. Reg. 10205, March 2, 1998. See also Armed Forces (2000).

[12]42 U.S.C. § 653.

[13]The Privacy Act requires the data collector to "establish appropriate administrative, technical, and physical safeguards to insure the security and confidentiality of records." Similarly, the Act requires the data collector "to protect against any anticipated threats or hazards to their security or integrity which could result in substantial harm, embarrassment, inconvenience, or unfairness to any individual on whom information is maintained."

[14]*Whalen v. Roe*, 429 U.S. 589 (1977).

a state-run, centralized, computerized database, the names and addresses of anyone who had obtained certain drugs pursuant to a doctor's prescription. The Supreme Court held that the New York database containing massive amounts of sensitive medical information passed constitutional muster. In particular, the Court cited New York's need for the database as part of the state's war against drugs and noted that the New York statute and its related implementation, including extensive database protections, showed "a proper concern with, and protection of, the individual's interest in privacy." The Court, however, reserved for another day consideration of legal questions that could arise from unauthorized disclosures of information from a government database "by a system that did not contain comparable security provisions."[15]

Concerns about misuse of data and the possibilities of identity theft could be addressed in part by Army procedures already established to respond to Privacy Act requirements. Specifically, the Privacy Act requires the Army to use appropriate data safeguards and carries the threat of civil and criminal sanctions if these requirements are not carried out.

In summary, from the perspective of those concerned about the Army's biometrics program and center, the Privacy Act addresses concerns related to the purpose for which the data will be collected, the notification of individuals that their personal information is needed, and how the information will be used and shared. In addition, the Privacy Act requires government agencies and officials to secure their database.

The Privacy Act permits many exceptions, and the Army could make many routine-use exceptions for biometric identification information. Although additional uses of the data beyond the original purpose for which it was collected must be published in the *Federal Register* and shown to be compatible with the original use, this is not a serious obstacle to the sharing of data. Thus, it leaves the individual without much firm protection against function creep.

The Army should strive to address function creep concerns through laws, regulations, and policies designed to ensure that biometric data

---

[15]*Whalen v. Roe*, 429 U.S. at 605–606.

are used only for purposes explicitly stated and that even marginal changes are reviewed by appropriate policymakers or senior leadership. Protection against function creep will help ensure that the data are not used for tracking unless that is a specifically stated goal.

OTHER MILITARY POLICIES ADDRESS SPECIFIC SOCIOCULTURAL CONCERNS

Beyond the issues raised by the Privacy Act, the Army can rely on other military policies and procedures to further address religious concerns. Precedents found in the case law address intrusiveness concerns.

Religious Objections

The military is no stranger to addressing the religious concerns of its members. The Army's use of biometrics is likely to be met with objection on religious grounds by a small number of personnel. The Army has an extensive, established policy in place to accommodate religious practices.[16] The Army approves requests for accommodation of religious practices unless the accommodation will have an adverse impact on "military necessity," which consists of unit and individual readiness, unit cohesion, morale, discipline, safety, and health. As the Army's primary advisor on matters pertaining to religious accommodation, the Army Chief of Chaplains is an important institutional asset on whom the Army leadership may call for guidance in determining how religious objections to biometric applications should be handled. As the official charged with establishing the Army's policy on the accommodation of religious practices, the Deputy Chief of Staff for Personnel (DCSPERS) will also have a key role to play.

Physical Privacy

The federal courts have yet to decide any cases involving an individual's refusal to participate in a biometric program mandated by the federal government. However, the Army can draw parallels to legal

---

[16]See Army Regulation (600-20, 1999).

challenges brought to fingerprinting in a noncriminal context—when an individual must provide a copy of his fingerprints for employment purposes, for example. The courts have generally upheld federal, state, and municipal requirements for fingerprinting as a condition of employment and licensing, provided the government has a rational basis for the requirement. For example, a union representing some 5,170 utility workers employed at nuclear power plants challenged as unconstitutional that part of a federal statute requiring these workers to be fingerprinted. The federal court disagreed with the union and upheld the fingerprint requirements. The court found that the U.S. government had a rational basis for requiring fingerprinting, namely concern over security at nuclear power plants.

What will happen if the Army wants a specific biometric from a person in a criminal justice context—because it suspects the person of a crime, for example? The Fourth Amendment to the U.S. Constitution governs searches and seizures conducted by government agents. As the amendment makes clear, the Constitution does not forbid all searches and seizures, only "unreasonable" ones. The Supreme Court defines a search as an invasion of a person's reasonable expectations of privacy.[17] To evaluate whether providing a biometric identifier in a criminal justice context constitutes a search, the judiciary focuses on two factors. First, the Court examines the nature of the intrusion.[18] Actual physical intrusions into the body, such as blood-drawing, breathalyzer testing, and urine analysis, can constitute Fourth Amendment searches. Second, the Court examines the scope of the intrusiveness, paying close attention to the "host of private medical facts" revealed during the search.

In cases where provision of a biometric identifier might be found to constitute a search (as in the hypothetical case of a physically intrusive DNA-based biometric that would reveal extensive medical facts), the ultimate measure of the constitutionality of a governmental search is "reasonableness."[19] To make this determination, a court must balance the "intrusion on the individual's Fourth Amendment interests against its promotion of legitimate governmental

---

[17]See, e.g., *Katz v. United States*, 389 U.S. 347, 360-62 (1967) (Harlan, concurring).

[18]See *Skinner v. Railway Labor Executives' Ass'n*, 489 U.S. 602, 616 (1989).

[19]*Vernonia Sch. Dist 47J v. Acton*, 515 U.S. 646, 652 (1995).

interests."[20] For a search to be reasonable, in the criminal justice context, the Army must generally show probable cause to believe that the person or place searched is implicated in the crime.

RESPONDING TO SOCIOCULTURAL CONCERNS WITHIN A BROADER APPROACH IS CRITICAL

Our legal assessment has raised no significant obstacles to the Army's establishment of a biometrics program or center in the United States. In fact, many military regulations and procedures are already in place to help address the concerns that might be raised by a biometrics program and center. By its implementation of the Privacy Act and its own policies on religious accommodation, the Army addresses two of the three major concerns identified with biometrics: concerns about privacy, including fears of government tracking and the collection of additional information or distribution of personal information, and infringement on religious freedoms. Concerns about misuse of data and identity theft are addressed, though less directly, through provisions in the Act related to the obligations of the data collector to protect security. Concerns about physical intrusiveness do not seem to have legal standing for current biometrics although this may change if medical information becomes part of the personal information revealed in a biometric.

These findings do not mean that the Army can avoid responding to perceptions about biometrics. Despite its legal feasibility, biometrics use by the Army could still be overturned by concern in the military or in the larger society about the program. Given these concerns, it is critical for the Army to take additional steps to improve perceptions of its biometrics program. Such an approach includes four steps.

- Thoroughly explain why biometrics are the best solution to a particular problem.
- Structure a program and select technologies to minimize effects on privacy.
- Educate the Army community and the public about the purpose and structure of the Army's program.

---

[20] *Id.* (quoting *Skinner*, 489 U.S. at 619) (internal quotation marks omitted).

- Assign responsibility within the Army for guiding these steps.

Thoroughly Explain Why Biometrics Are the Best Solution to a Particular Problem

The Army must be prepared to justify its program by defining a compelling problem and explaining why biometrics (or a specific biometric application) are the preferred solution. The justification should include a detailed description of the problem it intends to address with biometrics. It should describe and evaluate a range of alternative solutions, including biometrics. The criteria for evaluation should be clearly stated. This justification is the basis on which individuals and society will decide whether their concerns about biometrics should be secondary to the common good. Policymakers will base their judgments on these arguments as well.

Many individuals, particularly those in the military, have a high regard for the need to protect information and facilities, and those arguments, if backed by sound analysis, will carry considerable weight with soldiers, civilians, contractors, and families. To many of them, a well-thought-out biometrics program will be a logical improvement to existing security procedures. The threat posed by weaknesses in the current programs and the importance of biometrics to solving this problem will be the basis of decisions made by policymakers and legal counsel with regard to the structure of the program, the importance of enforcing compliance, and the extent of accommodations for legitimate individual concerns.

There are no laws, regulations, or specific procedures in the Army or DoD to ensure that the Army defines a compelling problem and that biometrics are the best solution. Were the Army to be the sole proponent of biometrics, this analysis would come from within as part of the argument for funding a biometrics program. Given pressures from outside DoD to move these activities forward, the problem analysis could be neglected—an oversight that could prove destructive as the program becomes more concrete.[21]

---

[21]Moreover, fast-paced commercial developments with the emerging technology could also push the Army in uncomfortable or unexpected ways. For example, if, in the near future, major computer manufacturers, software developers, and Internet

The justification will define the purpose of the biometric program. If the purpose is clearly related to a problem and is narrowly defined, individuals are likely to be less concerned about giving up information (and privacy) than if the purpose is more broad-based. The importance of ensuring a clearly defined purpose is supported by the Supreme Court's reasoning in its leading case on informational privacy, *Whalen v. Roe*. In *Whalen*, the Court found that the New York statute mandating a computerized database addressed a compelling problem and that the state agency had taken comprehensive measures to protect the database from unwarranted intrusions.

Defining the purpose also provides a context for determining the data to be collected. If the purpose of the program is to verify the identity of individuals for facility or information access control, there would be no need to collect information other than the biometric. If the Army wants to include other data, such as levels of access or a person's name, it should make the reasons for this clear to the participants and the general public. Because the biometrics currently in use have not been shown to contain medical information, the answer to the question, "What data is provided by the biometric?" would seem to be "None, other than the physical characteristics associated with an individual's biometric."

Structure a Program and Select Technologies to Minimize the Effects on Privacy

Three issues must be addressed in making decisions about how to design a biometrics program and select technologies:

- Policies about sharing data.
- Privacy enhancing solutions.
- Data repository choices.

---

providers embrace biometrics for computer and network access, this commercial development might influence Army commanders to demand the technology for immediate deployment. Instead of having in place a uniform policy for biometrics and their use, the Army might approach biometric use piecemeal, without any comprehensive planning in place. Although experimentation is generally welcome, the Army should have a biometric policy in place that will, at a minimum, address sociocultural concerns.

Policies about sharing data determine who will have access to the biometric database and for what purposes. From a privacy perspective, policies should be as restrictive as possible to limit the possibilities of function creep and the development of tracking capability. Even if the database contains only biometric information, concerns will arise about the extent to which the information will be shared to serve other purposes and whether and how participants will be notified of these additional purposes. For some purposes, however, other data, such as financial or medical information, might be tied or linked to the biometric.

As noted earlier, decisions the Army makes related to sharing of biometric identification information will be viewed as an important indicator of how seriously the Army takes its privacy protection responsibilities. As the Army considers which exceptions might be requested with respect to biometrics, it might benefit from study of the ongoing FBI experience involving the bureau's searchable criminal and civil fingerprint databases. In particular, the Army will be interested in the conclusion of the FBI's Office of General Counsel that the FBI's use for criminal justice purposes of fingerprint records obtained from service members and federal employees among others is legally unobjectionable. (See Appendix C, FBI Experience.) However, such a use will be a matter of concern to those who would like to limit sharing of personal information between military and law enforcement organizations.

Establishing a board or committee to assess biometric data policies, particularly with respect to data sharing for medical or biometrics-related research could also show a good faith effort to protect individuals' privacy. Policies regarding the destruction of biometric data will also affect privacy perceptions. Holding biometric data after a person loses access to a particular system or facility or leaves the service will exacerbate perceptions that the Army is collecting these data for purposes other than those stated.

Privacy enhancing solutions should be one of the criteria the Army uses in choosing biometric technologies. The design of data authentication and storage procedures will affect privacy perceptions. In choosing the biometric technology and structuring its program, the Army can address some of the privacy concerns raised about biometrics. Today, verification applications are thought to enhance privacy more than those using biometrics for identification. At a minimum, biometrics should add a level of protection and, in some cases, are likely to enhance privacy, such as applications using smart cards protected by biometrics (as in the Fort Sill pilot program).

Using less distinctive or robust biometrics for appropriate applications, such as verifying access by a limited group of authorized users, is another way to limit function creep and tracking capabilities. The INS's INSPASS program used at selected airports as an alternative to waiting in the immigration line is an example of this. Hand geometry is used because the relatively few users are prescreened. Yet hand geometry is probably not distinctive enough to use in an identification application.

Another way to address privacy concerns is to use biometrics with multiple options or "availability," such as fingers, and to use different biometrics for different programs. This would make it more difficult to engage in tracking and might limit pressure for function creep by reducing the number of templates in any particular technology.

Although use of multiple biometrics for different purposes in the Army might raise concerns that it is building a tracking database using many different features, some privacy experts would take a different view. Some have proposed that use of multiple biometrics protects an individual by ensuring compartmentation.[22] For example, a fingerprint used to gain access to the lab cannot be associated with a facial image used to get into the base exchange. All the biometric can do is ensure that the fingerprint or the image matches templates authorizing access to those facilities. The system does not need to know to whom these images belong or how to connect the purchasing patterns of the scientist with the hours he is in the lab.

Using multiple biometrics depending on the program needs and minimizing use of central repositories will not only help alleviate concerns about tracking, but it will also minimize demands on the Army to share the data, connect databases, and contribute to function creep. Technical obstacles to connecting data as well as policy and procedural practices that limit the purpose and uses of the data will bolster the Army's position that it is serious about the privacy of

---

[22]See, e.g., Wayman (1998).

biometric data. Commitments to avoid biometrics that contain medical information, such as DNA, even if they become commercially viable, might be another policy the Army would want to adopt.

The various experts interviewed repeatedly stressed that careful planning and attention to detail are important components of a successful biometrics program. Detailed planning includes such steps as addressing the accessibility of biometric systems for persons with disabilities, determining whether particular individuals in a facility will object to certain biometric technologies or would be unable to enroll, and realizing that military affiliated personnel and the public at large will have limited tolerance for large-scale biometric programs that do not work or work only for a subset of those enrolled. Planning should not be limited to the performance of the technology but must consider the range of people using the biometric, ensuring that physically impaired people can use the device.

Data repository choices also affect perceptions of privacy and security. One way to enhance security is to use biometrics that do not require templates to be sent to a central repository for matching or to decentralize storage and matching altogether, using a smart card. Systems that send a template to a central repository for comparison run the same risks as other information transmitted "over the wires." That is, the information transmitted can be intercepted at a number of points, resulting in the theft of either the biometric template or the authorization to accept the stolen or blocked biometric, depending on the purpose of the saboteur. Encryption, use of sequence numbers, time stamps, and other electronic data protection methods can help safeguard these transmissions, but they are not as inherently secure as systems that do not transfer the data. On the other hand, programs protecting access at one location, be it desktop or building or base, can design systems in which all the data is held locally.

Educate the Army Community and the Public About the Purpose and Structure of the Program

Openness and education about the program are two ways to address concerns that the biometric data contain additional information or that the data will be used inappropriately. Whether the program is large or small, personnel will need to be informed of the program and educated about it. The Army's education campaign should address the following questions:

- What is the purpose of the biometric program? Who is included in it?
- What information will be available through the biometric?
- How will that information be used and who will have access to it?
- How will that information be protected?
- Who will establish, control, and review these practices?

Although it is possible to undertake smaller programs with limited publicity, the establishment of a national biometric center would require considerable public education on these topics. Nearly all the program experts, lawyers, and ethicists interviewed for this project noted the importance of an education campaign to build support for any personal information collection program, particularly when a large population will be included. The Army's experience with biometrics includes small- as well as large-scale programs, such as those at Fort Sill or the military's DNA Human Remains Identification program. A critical component of their success has been educating the personnel involved about the purpose and limits of the program, as well as control of the data records.

Assigning Responsibility in the Army for Guiding These Steps

The Army has many decisions to make as it develops a larger biometrics program and, potentially, a biometrics data repository and test center. The issue of who guides these decisions initially and in the future is critical to the Army's ability to sustain its biometrics program. Who reviews requests for access to biometric data—be it for medical research or law enforcement purposes? Who ensures that the biometrics program is responsive to privacy concerns?

The Army's choices in implementing its biometrics program, and particularly when establishing a biometrics repository, should be consistent with the need to protect privacy and with its public commitments to do so. Decisions must be made about where to use biometrics, who must be included, and what data will be linked to the biometrics.

Chapter Five

WHAT IS THE FEASIBILITY OF A NATIONAL BIOMETRIC CENTER?

As noted in Chapter One, biometrics are a potential solution to Army needs. However, the findings of the previous two chapters indicate the Army must justify its program based on specific problems for which specific biometrics are an appropriate solution. As shown in Chapter Four, the feasibility of a biometrics program rests on how the program is structured and whether its implementation adequately addresses concerns about informational privacy, physical privacy, and religious objections as identified in Chapter Three. The establishment of a biometrics center, to include an RDT&E facility and a repository has a good chance of success if it is justified by program needs and structured and managed to provide the greatest privacy protection possible.[1]

The question for the Army is what kind of center makes sense for the Army to operate. A center could focus on RDT&E exclusively or include a repository where templates would be stored. Along these lines, the center's role could range from a very modest RDT&E facility with no template repository that would serve Army needs exclusively to a larger Army-operated center playing the lead role for DoD to a truly national center that would include RDT&E and a template repository that could be the focal point of all the U.S. government's biometrics work.

---

[1]In researching and writing this chapter, the authors relied heavily on the following sources: Newton and Rubenson (1999), Newton and Webb (1999), and Wayman (1999c).

Based on our analysis, we find that unless significant benefits to the Army are associated with running a national biometric repository, taking on this challenge, particularly at the beginning of the Army's foray into biometrics, runs the risk of raising concerns that could threaten the entire program. However, a biometrics RDT&E center, if warranted by Army needs, could be led by the Army without the sociological, legal, and ethical concerns related to the running of a repository.

In either case, education of the public is critical to success. The establishment of a center, whether for RDT&E or a data repository, is not a program to just be "slipped into" legislation. Particularly in the case of a large-scale data repository, the Army must make a compelling argument for managing such a repository. To do this, it must be able to provide convincing answers to the five questions raised in Chapter Four.

- What is the purpose of the biometric program? Who is included in it?
- What information will be available through the biometric?
- How will that information be used and who has access to it?
- How will that information be protected?
- Who will establish, control, and review these practices?

BIOMETRIC RDT&E CAPABILITIES

To evaluate potential RDT&E capabilities of an Army biometrics center, it is helpful to examine the role played by the National Biometric Test Center (NBTC) and the challenges found in testing biometrics. The U.S. government runs the NBTC at San Jose State University in California. James L. Wayman directs the NBTC, which is funded primarily through the NSA. NBTC's research interests are in application-specific testing of systems and in developing statistical methodologies for testing.[2]

---

[2]One of the reasons for NBTC's limited research agenda is that NSA is severely limited by law as to which data it can collect from U.S. citizens. See generally Executive Order 12333 (1981).

Commercial vendors and biometric consultants undertake evaluations of biometric devices. However, vendor testing alone fails to provide adequate information. From the perspective of motivation for testing, vendors and end-users have diverging goals. Vendors' reasons for testing include improving their devices and using the test results to sell their products. End-users seek testing results that will aid them in selecting a device that best fits their needs. Their focus is specific to their application and their enrollees.

Errors that affect biometric authentication devices potentially come from four different sources: variations in the biometric pattern, the presentation of the biometric to the sensor, the sensor, and the transmission process (including noise introduced by compression and expansion). Each of these factors is strongly tied to a specific application. One test environment cannot predict error rates for all applications. Hence, results from laboratory testing (vendor or otherwise) are highly dependent on the testing scenario population and are not necessarily a useful predictor of errors in real-world uses that differ from the testing scenario.[3]

Three important types of testing include algorithm verification, operational testing, and scenario evaluation:

- *Algorithm verification* occurs when testers evaluate algorithms used by a single device employing a database of "standard" samples. The results of this testing determine which algorithms are "good" and which are "poor." Although these tests are useful and repeatable, the results do not show real-life performance under real field conditions with real enrollee populations.
- *Operational testing* is typically used to evaluate pilot programs. It helps determine how the system will perform as a whole based on a specific application environment and the target population.
- *Scenario evaluation* is used to test the performance of multiple biometric systems in a modeled real-world application of interest to evaluate and compare performance across biometric devices. All devices are tested in the same environment on the

[3]Vendor or scientific laboratory testing generally presents only one scenario of biometric application: overt, cooperative, habituated, supervised, standard, closed, and private. See Appendix A.

> same population. This method allows for comparison of devices of different types. Scenario evaluation helps an end user decide which biometric device will work best for the end user's needs.[4]

What complicates testing is the fact that samples from thousands to millions of people are needed to test the distinctiveness of a biometric. Testing on these large sample sizes enables researchers to draw conclusions about uniqueness that are statistically significant. Biometrics also "age" or change over time. To acquire samples over any amount of time (from weeks to months or even longer) in any number of contexts from this number of people would be close to impossible, and to do this same testing for the many variables in each type of application would be even more difficult and probably too costly.

James Wayman (1999c) has summarized the three major difficulties in testing biometric devices and systems: "the dependence of measured error rates on the application classification, the need for a large test population that adequately models the target population, and the necessity for a time delay between enrollment and testing." Operational (e.g., field testing or pilot testing) and scenario evaluations, while expensive, are the only reasonable methods to test a system fully and reliably for deployment. Additionally, laboratory testing can be used to evaluate algorithms and as an initial pass/fail test for a biometric device to pass minimum standards before it is tested further either operationally or in a scenario evaluation.

An Army RDT&E center may also undertake further development of mathematical and statistical methods for test design and evaluation of biometric systems. It could also help develop standards or best practices for template collection, compression, and storage. An RDT&E center could be a source of advice on biometric systems for agencies inside and outside the Army, providing help to develop the educational roll-out piece of another agency's biometric program.

A CENTER FOR BIOMETRIC RDT&E SEEMS FEASIBLE

If the Army decides to pursue biometric technologies, it will likely want to establish a biometric RDT&E center to coordinate various

[4]Appendix A also discusses several other types of testing.

biometric activities, evaluate potential technologies, and adapt them to Army needs. As in other such labs, some work might be conducted in-house, but other functions would be contracted to other centers. Because the other armed services are also interested in the potential of biometrics, there may be calls for joining forces in a DoD-wide biometric RDT&E center. Similarly, other federal and state agencies working on biometrics might welcome the Army's help in creating a national RDT&E center that would serve broader governmental interests. For example, federal law enforcement agencies might eventually be interested in Army biometric data for use in criminal investigations. Furthermore, many federal agencies with whom the Army exchanges data might want to ensure that biometric data are standardized and systems are interoperable. In addition to test and evaluation functions, a larger center would presumably coordinate activities of the various members, ensure that technological developments and test results are shared, and even determine where future research efforts should be focused.

Whether the Army should seek the role of executive agent for a national biometric RDT&E center depends on the importance biometrics are expected to have in the Army. For example, if the success of the digitized Army depends on biometric technologies to protect battlefield intelligence nodes, then biometrics will be a high priority for the Army. If biometrics are one of many possible solutions to this problem, then perhaps the Army might want to retain more flexibility by pursuing biometrics within its own RDT&E structure or within a broader DoD structure.

AN ARMY OR DoD REPOSITORY FOR BIOMETRIC DATA ALSO SEEMS FEASIBLE

Based on our preliminary research, we believe an Army or DoD centralized repository for biometric data is feasible but not necessarily critical to the success of an RDT&E center. However, the feasibility ultimately depends on the purpose of the repository. This section details potential purposes for a repository and identifies the benefits and challenges associated with these purposes.

At a minimum, a central repository could be used to store templates collected for biometric programs. A central repository could play many useful roles. For example, if local authentication files are corrupted or erased, the central repository would have continuity of operations files or backup files that could be used. Furthermore, Army personnel move frequently and on short notice, so it might be useful to have a central repository for templates that could easily be transferred between facilities, much as security clearances are. However, given the ease of regenerating a template, the administrative difficulties of identifying the right template file, and concerns that old templates might not match the current individual (for biometrics that make adjustments with each use), it seems unlikely that the Army would rely solely on the repository to replace locally maintained templates.

Another reason to store templates is to support quality control and research at the biometric RDT&E center. Access to large numbers of biometric templates would help researchers test algorithms. However, templates would not be helpful in testing integrated systems because they would not provide such variables as user-sensor interactions and liveliness tests that are critical to the performance results of biometric systems.[5] Because template data can be transferred electronically, some might contend that no overwhelming need exists to co-locate the repository with an RDT&E center. The template data could be stored anywhere and simply accessed with a computer by those authorized to do so.

Finally, the Army may have legitimate reasons to store templates in a central repository. First, co-location with the RDT&E center could reduce risks associated with interception of biometric data. Second, at some point the Army could perform identification searches on the entire database of templates or on subsets of the database. This type of searching is likely to become much more feasible with advances in biometric technologies and computing power (see Appendix C, FBI Experience). Strengthened identification capabilities would allow biometrics to be used without accompanying cards or passwords, which would provide greater convenience to the biometric user.

---

[5]Liveliness tests are used in biometric applications to ensure that a person does not simply make a copy of your biometric, such as a Xerox of a fingerprint or a photograph of your face, and use this in the biometric sensor.

Concerns About a Centralized Repository

Security concerns about a centralized repository of biometric data are related to the vulnerability of that data as they are transmitted from the biometric reader to the repository and back again. The central repository might represent an attractive "honey pot"—attracting threats from hostile security services, hackers, and others seeking to compromise the integrity and security of network-accessible information. These vulnerabilities are not so different from those for passwords, PINs, or SSNs. Some would argue, however, that this vulnerability reduces the value of using biometrics, at least for remote verification applications. These transmission vulnerabilities can also be an argument for co-location of a test facility and repository, if they have an internal, highly protected network to safeguard the transfer of data.

A central biometric repository will also raise privacy concerns, depending on the purpose of the repository. Individuals will likely be more suspicious of a repository that only collects data than of one that verifies biometrics remotely. Given current computer power, widespread real-time matching against an entire DoD-wide repository seems unlikely to be feasible. However, as noted above, remote verification will raise privacy concerns, because the data are vulnerable to capture and tampering when they are in transmission. For some, storing all biometric data at one location can be more worrisome than having it dispersed because concentration of data lends credence to theories that the data will be used for tracking or purposes other than the ones advertised. A central repository would make sharing data with other organizations easier. At the same time, centralized control can make it easier to protect the data from misuse or function creep because only one staff member must be educated as to the rules for data protection and data-sharing.

Analysis

The issues raised here with respect to an Army biometric data repository are not likely to be significantly different if the repository served all of DoD. Considerable data-sharing exists at present and large amounts of personal data are consolidated in databases managed by the Defense Manpower Data Center, for example. Moreover, the military's emphasis on joint operations and other interservice endeavors suggests that the Pentagon might want to take a DoD-wide approach to biometric applications to ensure standardization and interoperability, rather than having each service field its own biometric systems.

Concerns about a repository are also likely to depend on whose biometric templates are included. Keeping service members' biometric data could account for millions of records, depending on the purpose of the program and the problem being tackled. As the circle is expanded to include Department of the Army or DoD civilian employees, contractors, retirees, dependents, and foreign nationals, more concerns will likely be raised about protection of privacy and purposes of the data and centralized repository. To the extent it is concerned about ensuring data privacy, the Army, by operating a repository serving its own or DoD's needs, could take the lead in influencing data protection and related policies.

If the Army takes the lead on a DoD repository, its role could expand to become that of the head of a national biometrics repository, serving all of the federal government, for example. It is not clear what the purpose of a national repository would be, and, thus, it is difficult to evaluate the feasibility of establishing such a national center and the value to the Army of running it.

A NATIONAL BIOMETRICS DATA REPOSITORY RAISES SERIOUS FEASIBILITY ISSUES

Based on our research and analysis to date, a national biometric data repository raises serious feasibility issues. While a national repository could serve useful purposes, such a center may provoke concerns for privacy protection.

Among the useful purposes it could serve, a national biometrics repository could provide for efficiency and interoperability. If other federal agencies using biometrics also find it necessary to store templates, and a large repository of data from military personnel already exists, then other federal agencies might want to build on the Army's expertise and use the Army's repository for template storage—perhaps on a fee-for-service basis. Congress might also proceed further down the path of data-sharing for law enforcement purposes and take steps to ensure that biometrics collected for military, federal employment, and licensing requirements are made available to law enforcement authorities under certain conditions. The FBI appears to be considering a move in this direction when automation of fingerprints makes this possible. Such data-sharing might also be envisioned as a method to reduce fraud in social service programs, in much the same way as states have used biometrics to identify double-dippers. Perhaps a national biometrics data repository would be made available to federal and state social service agencies to ensure that only qualified individuals are receiving benefits.

Opponents of such a national biometrics repository will draw comparisons to the ubiquitous use of SSNs as an identification number despite the original assurances by the government that the SSN was not to be used for other purposes. As the population providing biometrics for the repository grows, so too will objections increase. Servicemembers, who already sacrifice many of their freedoms to serve in the military, are likely to be compliant. Similarly, the large circle of people with ties to DoD will also be expected to generally accept the need for a centralized repository. However, expanding the circle of participants to include other federal employees, social service recipients, and people who have had background checks will increase the number of people likely to raise concerns.

While the Army might like to have access to others' biometric data, such as fingerprints, it is not clear that its need is sufficient to take on the burden of running a national repository. The Army can obtain this information in many ways, such as data-sharing arrangements with other federal and state agencies. For example, if law enforcement purposes are the primary motivation, it would seem that the FBI should administer the national repository. Similarly, if fraud prevention in social service programs is the primary aim, perhaps the Department of Health and Human Services should take the lead.

With a national repository, not only would the possibilities for data-sharing be greater but pressure from various agencies to gain access to others' data would also increase if the center resided in one location, even if in separate databases. Furthermore, it seems likely that the agency in charge of managing the biometric repository would have access to much more data than any one agency in the federal government currently has.

Chapter Six

CONCLUSIONS AND RECOMMENDATIONS

In this chapter, we draw on the discussion in the previous five chapters to present our conclusions and recommendations.

CONCLUSIONS

RAND identified three potential sociocultural concerns with biometrics related to informational privacy, physical privacy, and religious objections.

With respect to informational privacy, laws, such as the Privacy Act of 1974, and regulations, such as the Army Privacy Program, provide a baseline for protecting an individual's privacy interest in his personal information. The Army has flexibility to either accept these standards as the maximum privacy protection it will give to individuals or provide additional protections beyond the Act's requirements, particularly with respect to limiting the sharing of biometric data.

As for physical privacy, the courts have generally upheld federal, state, and municipal statutes requiring fingerprinting as a condition of employment and licensure. Accordingly, the Army appears to face no significant legal obstacle in this regard, although it might have practical concerns if biometrics make people so uncomfortable they avoid or sabotage the system or complain to their elected officials, for example.

With respect to religious concerns, the Army already has in place detailed regulations to address conflicts between an individual's religious practices and Army procedures.

Thus, although the use of biometrics raises sociocultural concerns, based on today's perceptions of biometrics these concerns present no serious obstacle to proceeding with an Army biometrics program, provided the Army addresses the concerns in accordance with existing laws and regulations.

How the Army addresses these concerns will have an impact on how well its biometrics program is received. Simply following the letter of the law may be sufficient for a narrowly defined biometrics program, such as one designed to protect battlefield intelligence nodes. However, should the Army determine that biometrics are a viable solution for Army, DoD, or national information assurance needs, greater efforts to allay concerns will become important. The larger the population included in the Army's biometrics program, the larger the likelihood that some will object on informational privacy, physical privacy, or religious grounds.

Finally, as an emerging technology, biometrics are changing rapidly. While biometrics do not currently provide medical information, this might change as new biometrics become commercially viable or researchers begin to test whether certain biometric templates could be somehow linked to particular medical conditions or diseases. For example, DNA analysis may become automated to the point that it becomes feasible to use a DNA-based biometric in the authentication applications discussed in this report. Because of DNA's close association with medical information, mandated participation in a DNA-based biometric program is likely to be controversial in the Army community and the general public.

Because biometrics are rapidly changing technologies, society's views of information technology will also develop. The impact on biometrics of increased societal experience with computers, cyberspace, electronic commerce, the digitized world, and the Internet could cut either way, reducing concerns about biometrics or heightening public interest in greater privacy protection. The Army could be forced to address these spillover effects in the future.

RECOMMENDATIONS

As this report was being prepared for final publication, Deputy Secretary of Defense Rudy de Leon issued a memorandum on December

27, 2000, consolidating oversight and management of biometric technology under the recently created DoD Biometrics Management Office (BMO). This memorandum also called for the formal establishment of a DoD Biometrics Fusion Center (BFC) under the BMO. The BFC's purpose is to acquire, test, evaluate, and integrate biometrics and to develop and implement storage methods for biometrics templates. The BFC is located in Bridgeport, West Virginia.

This memorandum derived from Public Law 106-246, signed by President Clinton on July 13, 2000, which included the following provision: "To ensure the availability of biometrics technologies in the Department of Defense, the Secretary of the Army shall be the Executive Agent to lead, consolidate, and coordinate all biometrics information assurance programs of the Department of Defense."[1]

As the DoD BMO and the Army, as executive agent, continue to assess biometrics, they must carefully consider the sociocultural concerns biometrics raise, along with technical, operational, security, bureaucratic, and administrative issues. They should specifically consider the following recommendations, which address how the Army (and the BMO) should implement its biometrics program as well as identifying issues that the Army and BMO should explore. We begin by focusing on implementation.

Incremental Implementation

Based on current Army use of biometrics, it is not clear that the establishment of a national RDT&E center or a biometric data repository is necessary. The Army's biometric interests will best be served by an incremental approach to building a biometrics program and establishing a data repository.

This incrementalism need not limit the Army to a few applications or participants. Rather, it suggests that the Army should take a purposive approach, defined as focusing its biometrics efforts on specific problems the Army wants to solve and on specific purposes the Army wants to achieve. A purposive approach suggests that the Army

[1]For more information about the DoD Biometrics Management Office, visit the DoD BMO Web page, available at http://www.c3i.osd.mil/biometrics/.

might want to try many biometric pilot programs and tests involving different biometric technologies to help it determine what works best for solving a particular problem and achieving a certain purpose. Along these lines, the Army would likely benefit from greater participation in the U.S. government's Biometric Consortium, the federal focal point for much biometric research and development.[2]

If this purposive approach generates the need for an RDT&E center or a central repository for biometric data, then the Army will be on firm ground as it moves to establish these activities whether for the Army only or for DoD as a whole. A decision designating the Army as the federal government's executive agent of a national biometric center should be made only after careful consideration. If made prematurely, such a decision is likely to raise more questions for the Army than the Army is prepared to address as well as tax its bureaucratic resources to answer the questions.

Finally, as the Army implements biometrics programs, it will want to proceed with additional care when expanding the programs to include foreign citizens, whether they are employees on U.S. bases overseas or allies fighting on our side. The sociological and legal issues raised in this context may be more complicated to address than those related to U.S. citizens.

Privacy Act Implications

It is not clear that the Army's broad interests in providing for the nation's defense are significantly enhanced by sharing biometric information in its charge with other agencies, even if the other agencies' uses are also in the national interest. We believe the Army can gain much and lose little by taking an approach that protects the privacy of the Army and DoD communities in their biometric data. As a starting point, the Army should place strict limits on the sharing of biometric data. These limits should go beyond the minimum protections of the Privacy Act. The Army's approach to sharing biometric data should be that such data should not be shared unless the sharing is directly related to the purpose for which the biometric was taken. If other agencies believe it imperative for them to have access

[2]See Appendix D for a discussion of the Biometric Consortium.

to the Army's data, they should make their case to Congress or the White House about why they should be able to access it.

Education

The Army's biometrics program—whether it includes a collection of small, discrete programs; an Army-managed RDT&E center; or a centralized data repository—should be discussed publicly. The Army must assure participants, policymakers, and the public that biometrics are necessary to the Army's needs and that the technology's benefits outweigh any individual costs.

An education campaign will be important to gaining support from participants in the Army community and protecting the Army's program from critics on the outside. A comprehensive threat analysis is critical to the education campaign because it will help the Army make its case as to why the Army needs to use biometrics. Additionally, the Army's education campaign must reflect a thoughtful data-sharing and safeguarding program. If the Army believes it needs to use biometrics on a large scale, then it must work to be a model for the rest of government. It cannot afford to be an example of how not to do a biometrics program.

Choosing Technologies

The Army should consider the sociocultural concerns identified in this report when it chooses biometric technologies and designs biometric system architectures. In some cases, the biometric choice and system architecture design, including decisions about where the Army will locate template databases, can be made with the objective of enhancing privacy. A common concern is that locating template databases in a central repository and frequently transferring data from the field increases the system's vulnerability to hackers. This potential vulnerability may be avoided in part by decentralizing template storage and matching. For example, the Army's biometrically protected smart card at Fort Sill, where the biometric measured could be found only on the card, provided privacy protection for the participants. Similarly, the use of multiple biometrics or biometric diversity protects privacy because it makes for compartmentation.

The Army's selection of a biometric will be purpose driven. However, the Army will be better off avoiding biometrics that also contain additional information unless the threat analysis demands this type of biometric. Thus, even if a DNA-based biometric becomes commercially viable, we would discourage the Army from deploying it because of the associated concerns it raises about genetic information. In sum, as the Army chooses technologies it will also want to consider their effect on privacy concerns and societal perceptions, as well as their benefits for addressing a particular threat.

Implementation Oversight

As the Army pursues biometrics, it might want to establish a board or committee to assist with implementation. Such a board may be useful to screen requests for the sharing of biometric information, to develop data-safeguarding and data-destruction policies, to track biometric use in the Army and DoD communities, and to provide other oversight as required. The Army should develop an institutional asset that monitors Army biometric programs, including pilot programs and experiments. Ensuring procedural consistency would also help address sociocultural concerns. The board would help the Army maintain awareness and attentiveness to new concerns that might be raised about biometrics.

Additional Issues

As discussed more fully in Appendix C, the European Union (EU) Data Protection Directive has the potential to raise issues affecting the Army's use of biometrics in EU member states, whether on U.S. military facilities or in different settings. The Army should explore further the directive's implications and continue to monitor the implementation of the directive's compliance scheme. In particular, the Army should monitor the implementation of the "safe harbor" principles, which provide U.S. organizations with a means of satisfying the directive's requirement for "adequate" privacy protection. The Army should also pay close attention to how the directive's various exceptions and exemptions are interpreted because the scope of these exceptions may affect U.S. interests.

The Army has operations in many foreign countries at any given time. From the international perspective, the Army must be mindful of the sociocultural concerns raised by Army use of biometrics in foreign settings. As Army biometrics applications move overseas, research on the likely sociocultural concerns of foreign countries and regional organizations will need to be done. However, the Army would find the results of such research more useful if it specifies the purpose of the biometric program, the candidate biometrics, the foreign location, and who the Army wants to participate (e.g., foreign military personnel and/or civilians).

Concerns have also been raised about the need for systematic assessment of the Army community to better understand its views on biometrics. Because such assessments depend so much on the specific problem the biometrics are being used to address—i.e., their purpose, methods of using and safeguarding data, and who is to be included in the system—we believe it would not be productive to poll the Army community until a specific application is in mind. However, as part of a purposive approach to biometrics, the Army should use sociological research technologies to test receptivity to biometrics. As biometrics are identified as solutions to particular problems, it would be more useful to engage focus groups to gauge how biometrics will be received and to help educate potential participants.

The Army might support research on whether biometric templates contain medical information, whether they might in the future, and whether the information would be provided inherently in the biometric (as with DNA) or by inference (as with retinal scans that show changes that a medical professional might further research and interpret).

Finally, we stress that all of this analysis depends on the Army's explanation of its problems and how biometrics can fix these problems. The Army's explanation must be more than just a statement that the Army needs improved access controls to enhance the security of its informational and physical assets. The Army must explain the weaknesses of the current systems, options to address these weaknesses, and how biometrics can solve the problems. Such an analysis is critical to providing the basis for the Army community, policymakers, and the public to determine whether concerns about biometrics are outweighed by the benefits they bring.

Appendix A

BIOMETRICS: A TECHNICAL PRIMER

This appendix expands on information presented in Chapter Two. It begins with a definition of biometrics and related terms, and then describes the steps in the biometric authentication process, reviews issues of template management and storage, and addresses testing issues. The appendix concludes with a brief review of mainstream biometric applications.[1]

DEFINITIONS

A biometric is any *measurable, robust, distinctive* physical characteristic or personal trait that can be used to *identify*, or *verify the claimed identity* of, an individual. Biometric authentication, in the context of this report, refers to automated methods of identifying, or verifying the identity of, a *living person*.

The italicized terms above require explanation. *Measurable* means that the characteristic or trait can be easily presented to a sensor and converted into a quantifiable, digital format. This allows for the automated matching process to occur in a matter of seconds.

The *robustness* of a biometric is a measure of the extent to which the characteristic or trait is subject to significant changes over time.

[1]This primer does not cover standards for interoperability or so-called "plug and play" applications because this subject is tangential to this RAND project. In researching and writing this appendix, the authors relied heavily on the following sources: Hawkes and Hefferman (1999); Newton and Rubenson (1999); and Wayman (1999c, 2000). See also Dunn (1998) and generally, Jain, Bolle, and Pankanti (1998).

These changes can occur as a result of age, injury, illness, occupational use, or chemical exposure. A highly robust biometric does not change significantly over time. A less robust biometric does. For example, the iris, which changes very little over a person's lifetime, is more robust than a voice.

Distinctiveness is a measure of the variations or differences in the biometric pattern among the general population. The higher the degree of distinctiveness, the more unique the identifier. The highest degree of distinctiveness implies a unique identifier. A low degree of distinctiveness indicates a biometric pattern found frequently in the general population. The iris and the retina have higher degrees of distinctiveness than hand or finger geometry. The application helps determine the degree of robustness and distinctiveness required.

Living person distinguishes biometric authentication from forensics, which does not involve real-time identification of a living individual.

IDENTIFICATION VERSUS VERIFICATION

Identification and verification differ significantly. With identification, the biometric system asks and attempts to answer the question, "Who is X?" In an *identification application,* the biometric device reads a sample and compares that sample against every template in the database. This is called a "one-to-many" search (1:N). The device will either make a match and subsequently identify the person or it will not make a match and not be able to identify the person.

Verification is when the biometric system asks and attempts to answer the question, "Is this X?" after the user claims to be X. In a *verification application,* the biometric device requires input from the user, at which time the user claims his identity via a password, token, or user name (or any combination of the three). This user input points the device to a template in the database. The device also requires a biometric sample from the user. It then compares the sample to or against the user-defined template. This is called a "one-to-one" search (1:1). The device will either find or fail to find a match between the two.

Identification applications require a highly robust and distinctive biometric, otherwise the error rates falsely matching and falsely nonmatching users' samples against templates cause security problems and inhibit convenience. Identification applications are common where the end-user wants to identify criminals (immigration, law enforcement, etc.) or other "wolves in sheep's clothing." Other types of applications may use a verification process.[2] In many ways, deciding whether to use identification or verification requires a trade-off: the end-user's needs for security versus convenience.

In sum, biometric authentication is used in two ways: to prove who you are or who you claim you are and to prove who you are not (e.g., to resolve a case of mistaken identity).

APPROACHES TO AUTHENTICATION

In general, there are three approaches to authentication. In order of most secure and convenient to least secure and convenient, they are as follows:

- Something you are—a biometric.
- Something you know—PIN, password.
- Something you have—key, token, card.

Any combination of these approaches further heightens security. Requiring all three for an application provides the highest form of security.[3]

THREE BASIC ELEMENTS TO ALL BIOMETRIC SYSTEMS

All biometric systems consist of three basic elements:

- Enrollment, or the process of collecting biometric samples from an individual, known as the enrollee, and the subsequent generation of his template.
- Templates, or the data representing the enrollee's biometric.

---

[2]See, e.g., Appendix B, Program Reports, Fort Sill Biometrically Protected Smart Card.

[3]Security also depends on other factors, such as the care taken to apply security measures properly, insofar as safeguarding tokens and passwords and ensuring that transmissions of biometric data are adequately protected.

- Matching, or the process of comparing a live biometric sample against one or many templates in the system's database.

Enrollment

Enrollment is the crucial first stage for biometric authentication because enrollment generates a template that will be used for all subsequent matching. Typically, the device takes three samples of the same biometric and averages them to produce an enrollment template. Enrollment is complicated by the dependence of the performance of many biometric systems on the users' familiarity with the biometric device because enrollment is usually the first time the user is exposed to the device.

Environmental conditions also affect enrollment. Enrollment should take place under conditions similar to those expected during the routine matching process. For example, if voice verification is used in an environment where there is background noise, the system's ability to match voices to enrolled templates depends on capturing these templates in the same environment.[4]

In addition to user and environmental issues, biometrics themselves change over time. Many biometric systems account for these changes by continuously averaging. Templates are averaged and updated each time the user attempts authentication.

Templates

As the data representing the enrollee's biometric, templates are created by the biometric device. The device uses a proprietary algorithm to extract "features" appropriate to that biometric from the enrollee's samples. Templates are only a record of distinguishing features, sometimes called minutiae points, of a person's biometric characteristic or trait. For example, templates are not an image or record of the actual fingerprint or voice.[5] In basic terms, templates

---

[4]The system's ability to match the sample to the enrolled template is sometimes referred to as the biometric's reliability.

[5]Image files of fingerprints may be of interest to the Army because of their law enforcement applications. In the case of fingerprints, the Army may want to keep

are numerical representations of key points taken from a person's body.

The template is usually small in terms of computer memory use, and this allows for quick processing, which is a hallmark of biometric authentication. The template must be stored somewhere so that subsequent templates, created when a user tries to access the system using a sensor, can be compared. Some biometric experts claim it is impossible to reverse-engineer, or recreate, a person's print or image from the biometric template.

Matching

Matching is the comparison of two templates, the template produced at the time of enrollment (or at previous sessions, if there is continuous updating) with the one produced "on the spot" as a user tries to gain access by providing a biometric via a sensor.

There are three ways a match can fail:

- Failure to enroll.
- False match.
- False nonmatch.

Failure to enroll (or acquire) is the failure of the technology to extract distinguishing features appropriate to that technology. For example, a small percentage of the population fails to enroll in fingerprint-based biometric authentication systems. Two reasons account for this failure: the individual's fingerprints are not distinctive enough to be picked up by the system, or the distinguishing characteristics of the individual's fingerprints have been altered because of the individual's age or occupation, e.g., an elderly bricklayer.

---

both electronic image files of the fingerprint as well as the biometric templates. The image files are too large to be used for biometric applications but would be useful for forensic purposes. Moreover, the Army might want to store image files to give it greater technical flexibility. For example, if the Army did not keep image files of enrollees, it might have to physically reenroll each individual if the Army decided to change to a different proprietary biometric system. Image files are also known as raw data or the *corpus*.

In addition, the possibility of a false match (FM) or a false nonmatch (FNM) exists. These two terms are frequently misnomered "false acceptance" and "false rejection," respectively, but these terms are application-dependent in meaning. FM and FNM are application-neutral terms to describe the matching process between a live sample and a biometric template.

A false match occurs when a sample is incorrectly matched to a template in the database (i.e., an imposter is accepted). A false nonmatch occurs when a sample is incorrectly not matched to a truly matching template in the database (i.e., a legitimate match is denied). Rates for FM and FNM are calculated and used to make tradeoffs between security and convenience. For example, a heavy security emphasis errs on the side of denying legitimate matches and does not tolerate acceptance of imposters. A heavy emphasis on user convenience results in little tolerance for denying legitimate matches but will tolerate some acceptance of imposters.

TEMPLATE MANAGEMENT—STORAGE AND SECURITY

Template management is critically linked to privacy, security, and convenience issues. All biometric authentication systems face a common issue: Biometric templates must be stored somewhere. Templates must be protected to prevent identity fraud and to protect the privacy of users. A major concern is what additional information will be stored about each user along with his biometric template.

Possible locations for template storage include

- the biometric device itself,
- a central computer that is remotely accessed,
- a plastic card or token via a bar code or magnetic stripe,
- Radio Frequency Identification Device cards and tags,
- optical memory cards,
- Personal Computer Memory Card International Association cards, and
- smart cards.

In general, transmitting biometric data over communications lines reduces system security because the data become vulnerable to the same interception or tampering possible when any data is sent "over the wire." Biometrics are more secure when stored under the control of the authorized user, such as on a smart card, and used in verification applications.

Smart cards are the size of credit cards and have a microchip or microprocessor chip embedded in them. The chip stores electronic data that can be protected using biometrics. There are two types of smart cards: contact and contactless smart cards. A contact smart card must be inserted into a smart card reader to be used. A contactless smart card only has to be placed near an antenna to carry out a transaction.[6]

Another security issue for template database storage is whether the database will have a unique use or if it will be used for multiple security uses. For example, a facilities manager might use a fingerprint reader for physical access control to the building. The manager might also want to use the same fingerprint template database for his employees to access their computer network. Should the manager use separate databases for these different uses, or is he willing to risk accessing employee fingerprints from a remote location for multiple purposes?

In general, verification applications provide more security than identification applications because a biometric and at least one other piece of input (e.g., PIN, password, token, user name) are required to match a template. Verification provides a user with more control over his data and over the process when the template is stored only on a card. That is, such a system would not allow for clandestine, or involuntary, capture of biometric data because the individual would know if he were providing the card. Because the search only seeks a match against one template in the database, verification applications require less processing time and memory. Thus, they are less expensive than identification applications.

Additional security features can be incorporated into biometric systems to detect a "wolf," or unauthorized user. For example, a

---

[6]For a detailed discussion of smart cards, see Ratha and Bolle (1999).

"liveliness test" is a method of measuring if the biometric sample is being read from a live person versus a faux body part or body part of a dead person. Liveliness tests are done in many ways. The device can look for such things as heat, heartbeat, or electrical capacitance.[7] Other security features include encryption of biometric data and the use of sequence numbers in template transmission. A template with such a number out of sequence suggests unauthorized use.

MAINSTREAM BIOMETRICS AND THEIR APPLICATIONS

While there are many possible biometrics, at least eight mainstream biometric authentication technologies have been deployed or pilot-tested in applications in the public and private sectors.[8] These are

- fingerprint,
- hand/finger geometry,
- facial recognition,
- voice recognition,
- iris scan,
- retinal scan,
- dynamic signature verification, and
- keystroke dynamics.

Fingerprint

The fingerprint biometric is an automated digital version of the old ink-and-paper method used for more than a century for identification, primarily by law enforcement agencies. The biometric device involves users placing their finger on a platen for the print to be read. The minutiae are then extracted by the vendor's algorithm, which

---

[7]Electrical capacitance has proved to be the best and least reproducible method for effectively identifying a live person.

[8]For a detailed discussion of these mainstream biometrics, see Jain, Bolle, and Pankanti (1999).

also makes a fingerprint pattern analysis. Fingerprint template sizes are typically 50 to 1,000 bytes.

Fingerprint biometrics currently have three main application arenas: large-scale Automated Finger Imaging Systems (AFIS) generally used for law enforcement purposes, fraud prevention in entitlement programs, and physical and computer access.

Hand/Finger Geometry

Hand or finger geometry is an automated measurement of many dimensions of the hand and fingers. Neither of these methods takes actual prints of the palm or fingers. Only the spatial geometry is examined as the user puts his hand on the sensor's surface and uses guiding poles between the fingers to properly place the hand and initiate the reading. Hand geometry templates are typically 9 bytes, and finger geometry templates are 20 to 25 bytes. Finger geometry usually measures two or three fingers. During the 1996 Summer Olympics, hand geometry secured the athlete's dormitories at Georgia Tech. Hand geometry is a well-developed technology that has been thoroughly field-tested and is easily accepted by users.

Facial Recognition

Facial recognition records the spatial geometry of distinguishing features of the face. Different vendors use different methods of facial recognition, however, all focus on measures of key features. Facial recognition templates are typically 83 to 1,000 bytes. Facial recognition technologies can encounter performance problems stemming from such factors as noncooperative behavior of the user, lighting, and other environmental variables. Facial recognition has been used in projects to identify card counters in casinos, shoplifters in stores, criminals in targeted urban areas, and terrorists overseas.

Voice Recognition

Voice or speaker recognition uses vocal characteristics to identify individuals using a pass-phrase. Voice recognition can be affected by such environmental factors as background noise. Additionally it is unclear whether the technologies actually recognize the voice or just the pronunciation of the pass-phrase (password) used. This technology has been the focus of considerable efforts on the part of the telecommunications industry and NSA, which continue to work on improving reliability. A telephone or microphone can serve as a sensor, which makes it a relatively cheap and easily deployable technology.

Iris Scan

Iris scanning measures the iris pattern in the colored part of the eye, although the iris color has nothing to do with the biometric. Iris patterns are formed randomly. As a result, the iris patterns in your left and right eyes are different, and so are the iris patterns of identical twins. Iris scan templates are typically around 256 bytes. Iris scanning can be used quickly for both identification and verification applications because of its large number of degrees of freedom. Current pilot programs and applications include ATMs ("Eye-TMs"), grocery stores (for checking out), and the Charlotte/Douglas International Airport (physical access). During the Winter Olympics in Nagano, Japan, an iris scanning identification system controlled access to the rifles used in the biathlon.

Retinal Scan

Retinal scans measure the blood vessel patterns in the back of the eye. Retinal scan templates are typically 40 to 96 bytes. Because users perceive the technology to be somewhat intrusive, retinal scanning has not gained popularity with end-users. The device involves a light source shined into the eye of a user who must be standing very still within inches of the device. Because the retina can change with certain medical conditions, such as pregnancy, high blood pressure, and AIDS, this biometric might have the potential to reveal more information than just an individual's identity.

Dynamic Signature Verification

Dynamic signature verification is an automated method of examining an individual's signature. This technology examines such dynamics as speed, direction, and pressure of writing; the time that the stylus is in and out of contact with the "paper"; the total time taken to make the signature; and where the stylus is raised from and lowered onto the "paper." Dynamic signature verification templates are typically 50 to 300 bytes.

Keystroke Dynamics

Keystroke dynamics is an automated method of examining an individual's keystrokes on a keyboard. This technology examines such dynamics as speed and pressure, the total time of typing a particular password, and the time a user takes between hitting certain keys. This technology's algorithms are still being developed to improve robustness and distinctiveness. One potentially useful application that may emerge is computer access, where this biometric could be used to verify the computer user's identity continuously.

BIOMETRIC APPLICATIONS

Most biometric applications fall into one of nine general categories:

- Financial services (e.g., ATMs and kiosks).
- Immigration and border control (e.g., points of entry, precleared frequent travelers, passport and visa issuance, asylum cases).
- Social services (e.g., fraud prevention in entitlement programs).
- Health care (e.g., security measure for privacy of medical records).
- Physical access control (e.g., institutional, government, and residential).
- Time and attendance (e.g., replacement of time punchcard).
- Computer security (e.g., personal computer access, network access, Internet use, e-commerce, e-mail, encryption).
- Telecommunications (e.g., mobile phones, call center technology, phone cards, televised shopping).
- Law enforcement (e.g., criminal investigation, national ID, driver's license, correctional institutions/prisons, home confinement, smart gun).

Classifying Biometric Applications

Biometric applications may be classified in many different ways. James Wayman of the National Biometric Test Center suggests the following seven categories for classifying biometric applications, explained below.

1. overt or clandestine
2. cooperative or noncooperative
3. habituated or nonhabituated
4. supervised or nonsupervised
5. standard or nonstandard environment
6. closed or open system
7. public or private.

Overt versus clandestine capture of a biometric sample refers to the user's awareness that he is participating in biometric authentication.[9] Facial recognition is an example of a biometric that can be used for clandestine identification of individuals. Most uses of biometrics are overt, because users' active participation ensures performance and lower error rates. Verification applications are nearly always overt.

Cooperative versus noncooperative applications refer to the behavior that is in the best interest of the "wolf." Is it in the interest of "wolves" to match or to not match a template in the database? Which is to the "wolf's" benefit? This is important in planning a security system with biometrics. No perfect biometric system exists; every system can be tricked into falsely not matching one's sample and template—some more easily than others. It is also possible to trick a biometric device into falsely matching your sample against a template, but it could be argued that this requires more work and a sophisticated hacker to make a model of the biometric sample. One way to strengthen security in a cooperative application is to require a password or token along with a biometric, so that the "wolf" must

[9]James Wayman used "covert" instead of "clandestine."

match one specific template and is not allowed to exploit the entire database for his gain.

To gain access to a computer, a "wolf" would want to be cooperative. To attempt to foil an INS database consisting of illegal border crossing recidivists, a "wolf" (recidivist) would be noncooperative.

Habituated versus nonhabituated use of a biometric system refers to how often the users interface with the biometric device. This is significant because the user's familiarity with the device affects its performance. Depending on which type of application is chosen, the end-user may need to utilize a biometric that is highly robust. As examples, use of fingerprints for computer or network access is a habituated use; use of fingerprints on a driver's license, which is updated after several years, is a nonhabituated use. Even "habituated" applications are "nonhabituated" during their first week or so of operation or until the users adjust to using the system.

Supervised versus nonsupervised applications refer to whether supervision (e.g., a security officer) is a resource available to the end-user's security system. Do users need to be instructed on how to use the device (many new users or nonhabituated users) or to be supervised to ensure they are being properly sampled (such as border crossing situations with the problem of recidivists or other noncooperative applications)? Or is the application made for increased convenience, such as at an ATM? The process of enrollment nearly always requires supervision.

Standard versus nonstandard environments are generally a dichotomy between indoors versus outdoors. A standard environment is optimal for a biometric system and matching performance. A nonstandard environment may present variables that would create false nonmatches. For example, a facial recognition template depends, in part, on the lighting conditions when the "picture" (image) was taken. The variable lighting outdoors can cause false nonmatches. Some indoor situations may also be considered nonstandard environments.

Closed versus open systems refers to the number of uses of the template database now or potential uses in the future. Will the database have a unique use (closed), or will it be used for multiple security measures (open)? For example, a facilities manager might have his employees use a fingerprint reader to enter a building. He might also want to use the same fingerprint template database for employees to log on to their computer network. Should they use separate databases for these different uses, or do they want to risk remotely accessing employee fingerprints for multiple purposes? Other examples are state driver's licenses and entitlement programs. A state may want to communicate with other states or other programs within the same state to eliminate fraud. This would be an open system, in which standard formats of data and compression would be required to exchange and compare information.

Public or private applications refer to the users and their relationship to system management. Examples of users of public applications include customers and entitlement recipients. Users of private applications include employees of business or government. The users' attitudes toward biometric devices and management's approach will vary depending on whether the application is public or private. Once again, users' attitudes toward the device will affect the performance of the biometric system.

It should be noted here that performance figures and error rates from vendor testing are unreliable for many reasons. Part of the problem is that to test the distinctiveness of a biometric, anywhere from thousands to millions of people are needed to test theories of how "unique" a particular identifier is. To acquire samples over any amount of time in any number of contexts from this number of people would be impossible, and to do this same testing for the many variables in each type of application is in most cases impossible and in the others too costly if it were possible. Operational and pilot testing is the only reasonable method to test a system. Additionally, vendor and scientific laboratory testing generally present only one scenario of biometric application: overt, cooperative, habituated, supervised, standard, closed, and private (Newton and Webb, 1999).

SALIENT CHARACTERISTICS OF MAINSTREAM BIOMETRICS

Table A.1 compares the eight mainstream biometrics in terms of a number of characteristics, ranging from how robust and distinctive

Table A.1

Comparison of Mainstream Biometrics[10]

Biometric	Identify versus Verify	How Robust	How Distinctive	How Intrusive
Fingerprint	Either	Moderate	High	Touching
Hand/Finger Geometry	Verify	Moderate	Low	Touching
Facial Recognition	Either	Moderate	Moderate	12+ inches
Voice Recognition	Verify	Moderate	Low	Remote
Iris Scan	Either	High	High	12+ inches
Retinal Scan	Either	High	High	1–2 inches
Dynamic Signature Verification	Verify	Low	Moderate	Touching
Keystroke Dynamics	Verify	Low	Low	Touching

they are to what they can be used for (i.e., identification or verification or verification alone) (Newton and Webb, 1999). This table is an attempt to assist the reader in categorizing biometrics along important dimensions. Because this industry is still working to establish comprehensive standards and the technology is changing rapidly, however, it is difficult to make assessments with which everyone would agree. The table represents an assessment based on discussions with technologists, vendors, and program managers. The table is not intended to be an aid to those in the market for biometrics, rather it is a guide for the uninitiated.

When comparing ways of using biometrics, half can be used for either identification or verification, and the rest can only be used for verification. In particular, hand geometry has only been used for verification applications, such as physical access control and time and attendance verification. In addition, voice recognition, because of the need for enrollment and matching using a pass-phrase, is typically used for verification only.

There is considerable variability in terms of robustness and distinctiveness. Fingerprinting is moderately robust, and, although it is

[10]The authors compiled Table A.1 from various sources at the SJB Biometrics 99 Workshop, November 9–11, 1999, including Hawkes and Hefferman (1999). See also Jain, Bolle, and Pankanti (1998).

distinctive, a small percentage of the population has unusable prints, usually because of age, genetics, injury, occupation, exposure to chemicals, or other occupational hazards. Hand/finger geometry is moderate on the distinctiveness scale, but it is not very robust, while facial recognition is neither highly robust nor distinctive. As for voice recognition, assuming the voice and not the pronunciation is what is being measured, this biometric is moderately robust and distinctive. Iris scans are both highly robust because they are not highly susceptible to day-to-day changes or damages and distinctive because they are randomly formed. Retinal scans are fairly robust and very distinctive. Finally, neither dynamic signature verification nor keystroke dynamics are particularly robust or distinctive.

As the table shows, the biometrics vary in terms of how intrusive they are, ranging from those biometrics that require touching to others that can recognize an individual from a distance.

BIOMETRIC RDT&E CAPABILITIES

Biometrics are an emerging technology in an emerging industry that does not yet have comprehensive standards. As a result, test and evaluation will be an important component of an Army biometrics program.

The U.S. NBTC at San Jose State University in California, researches application-specific testing of systems and develops statistical methodologies for such operational and scenario testing. NBTC, directed by James L. Wayman, is primarily funded through the NSA. NBTC does not research or test specific biometric products. Its work must be linked to an application.

Commercial vendors also evaluate biometric devices. However, vendor testing is not independent, and results are not always replicable by others, such as the NBTC or the National Physical Laboratory (NPL) in England. In general, the performance of systems tested in a lab declines when the system is field tested.

There are six basic types of testing for biometric systems:

1. algorithm verification,
2. operational,

3. scenario,
4. usability,
5. security, and
6. template quality.[11]

NBTC Director Wayman and other experts include a seventh type of testing that is more cognitive: development of mathematical and statistical methods for test design and evaluation of biometric systems. Testing related to types 1, 2, and 3 are measurements of error rates.

Each basic type of testing is discussed below.

Test 1. Testers evaluate algorithms used by a single device using a database of "standard" samples. Standard samples are neither too easy nor too difficult for matching, and they probably do not reflect the anomalies found in populations. The results of this testing determine which algorithms are "good" and which are "poor." Although these test are useful and repeatable, the results do not show real-life performance under real field conditions with real enrollee populations.

Test 2. Operational testing is typically used for evaluating pilot programs. It helps determine how the system will perform as a whole based on a specific application environment on the target population.

Test 3. Scenario evaluation is used to test the performance of multiple biometric systems in a modeled real-world application of interest to evaluate and compare performance across biometric devices. All devices are tested in the same environment on the same population. The results are repeatable if the modeled scenario experiment can be controlled and should show real-life performance if done accurately. This method of evaluation allows for comparison of devices of different types. Scenario evaluation can help end-users decide which specific biometric device will work best for their needs.

---

[11]The discussion of the six types draws from information provided by Tony Mansfield, NPL, during a November 1999 interview and presentation in London, England (Newton and Webb, 1999).

Test 4. Usability of a biometric is critical to success, especially in commercial applications, because users must be willing to participate in the system and because usability enhances performance. Usability evaluations seek answers to such questions as whether the device is user-friendly, what difficulties users have with the system (e.g., intrusiveness, correct placement of biometric on the sensor), how the users' difficulties can be overcome, how users' difficulties with the system affect performance (e.g., false nonmatch rate), and whether the system is acceptable to end-users.

Test 5. Security evaluations seek answers to such questions as whether the system can detect imposters (a sufficiently low false match rate, liveliness tests, other measures), whether or not the device can distinguish between lookalikes (e.g., twins), whether some templates are easy to crack (i.e., does the device form templates for indistinguishable features that are easy to duplicate?), where the system is vulnerable, whether the system can be bypassed or hacked into and if so where and how.

Test 6. Evaluation of image/template quality could be useful for making standards for images'/templates' maximum allowable distortion, resolution, and signal-to-noise ratio. Standards for template quality will foster the ability for data-sharing between system managers. Through technical evaluation of biometric technologies, engineers and scientists search for the measurement of the following parameters: false match rate (FMR, the rate that a sample is incorrectly matched to a template in the database), false nonmatch rate (FNMR, the rate that a sample is incorrectly not matched to a template in the database), percentage of false nonmatches stemming from inconsistencies in the partitioning process[12] (known as the binning error rate), percentage of the total database to be scanned on average for each search (known as the penetration coefficient), transaction time (for finding resultant match or nonmatch), and failure to enroll/acquire percentage (percentage of the general population for whom the technology will fail to extract distinguishing features).

---

[12]Partitioning templates into smaller groups increases searching efficiencies and is used in systems holding a large number of templates. Partitioning can be based on information contained within the biometric template or other information gathered at the time of enrollment, such as the user's name or gender.

Even though biometric systems vary greatly across both biometric type and vendor, biometric devices have five subsystems: data collection, transmission, signal processing, storage of templates, and decision. Data collection occurs at the human-machine (sensor) interface and includes the creation of the biometric template. Transmission refers to the communication of the biometric template between the sensor and the next subsystem (either signal processing or storage depending on whether the user is trying to match a template[s] or enroll). This may include compression and subsequent expansion, which may add noise to the biometric pattern. Signal processing is when the device extracts distinguishing features from the biometric pattern presented at the sensor for matching and compares it to the template(s) stored during enrollment. Storage of a biometric template occurs at enrollment. The decision process takes the score from the signal processing subsystem and decides if a match or nonmatch is found, based on the thresholds the end-user has put into the system.

One test environment cannot predict error rates for all applications. Errors that would affect biometric authentication devices potentially come from four different sources: variations in the biometric pattern, the presentation of the biometric to the sensor, the sensor, and the transmission process (including compression and expansion noise). Each of these factors is strongly tied to a specific application. Hence, results from laboratory testing (vendor or otherwise) are dependent on the testing scenario and cannot usefully predict errors in real-world uses that are different.

Because vendor and scientific laboratory testing generally presents only the overt, cooperative, habituated, supervised, standard, closed, and private scenario, it is impossible to extrapolate performance in different sets of circumstances—such as in a nonhabituated or nonsupervised programs.

To test the distinctiveness of a biometric, anywhere from thousands to millions of people are needed to test theories of how "unique" a particular identifier is and to make statistically significant conclusions about uniqueness. Biometrics also "age" or change over time. To acquire samples over any amount of time (from weeks to months or even longer) in any number of contexts from this number of people would be close to impossible, and to do this same testing for the many variables in each type of application would be even more difficult and probably financially prohibitive.

To summarize, as James Wayman has explained, three major difficulties occur in testing biometric devices and systems: "[1] the dependence of measured error rates on the application classification, [2] the need for a large test population [that] adequately models the target population, and [3] the necessity for a time delay between enrollment and testing." (Wayman, 1999e.)

While expensive, operational field or pilot testing and scenario evaluations are the only reasonable methods to test a system for deployment fully and reliably. Laboratory testing could be used to evaluate algorithms on an initial pass/fail basis for a biometric device to pass minimum standards to be further tested operationally. An R&D lab may also undertake further development of mathematical and statistical methods for test design and evaluation of biometric systems. An RDT&E center could be a source of advice on biometric systems for agencies internal and external to the Army, including being the developers of the educational roll-out piece of a biometric program.

An RDT&E center will face the testing difficulties highlighted above. At the same time, it could be useful in targeting research, developing mathematical and statistical methods for test design and evaluation, and screening the algorithms initially.

Appendix B

PROGRAM REPORTS

The use of biometrics is increasing throughout the United States and the rest of the world. This appendix presents brief case studies of various public and private-sector entities employing biometrics to control access to facilities and computers, to prevent fraud, and to increase customer services, among other purposes.[1] The case studies pay special attention to privacy concerns and technology glitches (if any) that the Army should consider before a biometrics program can be widely implemented. This appendix concludes with case studies of other identifiers to include the DoD DNA specimen repository and the use of the SSN.

MILITARY PROGRAMS

Fort Sill Pilot Program: Biometrically Protected Smart Card

Problem: The Army sends recruits to basic training at one of five bases in the United States: Fort Sill, Oklahoma, Fort Jackson, South Carolina, Fort Leonard Wood, Missouri, Fort Knox, Kentucky, and Fort Benning, Georgia. Shortly after arrival at the base, the new recruits must buy toiletries, haircuts, and other personal items. To enable them to make these purchases, the Army issues recruits an advance on their pay. Giving these recruits several hundred dollars in cash causes concern because the money is easily lost or stolen. Thus, Fort Sill used a voucher system, while at Fort Knox, the Army

---

[1]For a description of some biometric applications, see, e.g., Gugliotta (1999); Hansell (1997); Rogers (various); and Mintie (various).

issued checks to the recruits and then marched them to the PX to buy money orders. These activities took hours to complete and complicated the training schedule. The Army's Training and Doctrine Command (TRADOC) and Finance Command began to look for alternative approaches. Because the Treasury Department's Financial Management Service and DoD's Defense Finance and Accounting Service (DFAS) manage government payments, they became involved in the search for solutions. The Army and Air Force Exchange Service (AAFES) also participated because it wanted to test speeding of throughputs and reduce cash handling at basic training points of sale.

Given the objective of a quick, safe, and efficient system of paying recruits, the Army decided to test three different systems of stored value cards containing digital cash:

- Smart cards that were PIN protected.
- Smart cards that were biometrically protected.
- Smart cards that were open purses, like cash.

The Army tested the biometrically protected smart cart at Fort Sill, Oklahoma (Moore, 1998). The biometric used was a fingerprint. Mellon Bank did the system integration. Identicator Technologies provided the biometric technology, with additional integration to the electronic purse done by Product Technologies, Inc.

Program: The biometric smart card pilot program began at Fort Sill in March 1998 and ran for 15 months. Determining the population to be included in the Fort Sill pilot was very straightforward, it would include all recruits arriving for basic training. Because Army basic training is a highly controlled environment, recruits have a very limited number of places at which they are allowed to spend money. The Army placed a smart card reader and fingerprint sensor at each location (points of sale) where recruits were allowed to spend money. The Army also gave a keychain-sized card reader to each drill sergeant to allow him to monitor how much money a recruit had on his card.

The first thing done with recruits arriving at Fort Sill was to verify their identity and SSNs and issue them the smart card with an advance on their pay. Army personnel enrolled each recruit into the system using a laptop computer and sensor to scan the recruit's right index finger to obtain a digital representation of the print. The clerk also scanned the recruit's left index finger as a backup. The clerk then added cash value to the card based on an Army formula: $200 for men; $260 for women. Cards were set to expire in 60 days at which time all remaining cash transferred to the recruit's bank account, which had been established in the meantime. The Army did not keep a separate record of the fingerprint; only the serial number of the card was linked to the cardholder name.

The recruit was responsible for keeping track of his smart card, which contained his fingerprint template. At points of sale, the recruit entered his card into the card reader and placed his right or left index finger on the sensor. This template was compared to the template on the card. If they matched, the sale went through with the amount deducted from the card.

Performance: The program team had no reports of fraud and no complaints about failure to use the system. Only 10 people out of the 25,000 enrolled during the pilot program were not able to enroll. This failure rate is much lower than the advertised 1 percent for fingerprint technologies. However, these young recruits are prime candidates for fingerprint biometrics. Of those enrolled in the system, only about 3 percent failed to gain access to their card with a first fingerprint, but, when a second fingerprint was used, there was 100 percent access. The only performance issue for the system was that after several months the sensors for enrollment would wear out and "go bad." The clerks managing the system learned to recognize the signs in advance and replace the sensors as necessary.

Protections: Drill sergeants are very protective of their recruits. During informational sessions with the project team, one drill sergeant expressed concerns that recruits with fundamentalist religious beliefs might object to using the biometric on religious grounds. They were also concerned that fingerprints would be kept after recruits left and were relieved to find out that no fingerprints from this activity would be retained by the Army. Even though sales were linked by serial number to a bank, and account information and names could be drawn from this, no information was gathered about recruit purchases. The drill sergeants were also assured that the template on the card could not be reverse-engineered into a fingerprint image. Because this pilot program did not involve personal information contained in a system of records, it had no Privacy Act implications. At the time of enrollment, each recruit received a brochure explaining the fingerprint technology.

No formal feedback was obtained from the recruits, but reportedly the drill sergeants found the program a great improvement over previous practices. The drill sergeants preferred the smart cards because of reduced risk of theft, which can be a time-consuming problem for them because they must assist with investigations, file reports, and help the recruit who has lost his money. However, the Army decided that while digital cash was a good solution, biometrics would not be used to protect the card. Experiments with open purse smart cards worked as well as those protected by a PIN or biometric but at less cost.

Lessons Learned: The program, directed from the top, was universally accepted without much difficulty. Several reasons explain this success. First, the program provided more time for training by reducing the time spent on administrative tasks. The program reduced from hours to minutes the process of paying recruits and conducting subsequent transactions for sundries, haircuts, etc. In addition, a well-thought-out educational campaign was targeted at the drill sergeants, the Army personnel who have the recruits' interests most at heart and the recruits' lives most in sight. Program managers showed some 300 drill sergeants how the technology worked, explained the limits to the information provided, and answered the sergeants' questions. Finally, the Army conducted the biometrically protected smart card program for a clearly defined purpose in a highly controlled environment with an ideal population.

All DoD training bases are now using open purse smart cards, except for the Marines at Parris Island, South Carolina, who use PINs to secure their cards.

Defense Manpower Data Center (DMDC)

The DMDC operates what is arguably DoD's largest biometric database. By way of background, the Federal Managers' Financial Integrity Act of 1982 requires federal managers to establish internal controls to provide reasonable assurance that funds, property, and other assets are protected against fraud or other unlawful use. As a result of this legislation, DoD launched Operation Mongoose, a fraud prevention and detection initiative. Operation Mongoose exposed a number of fraud schemes and indicated that DoD needed to improve servicemember identification and verification procedures. Responding to the need for better fraud prevention measures, the acting Under Secretary of Defense (Personnel and Readiness) gave authority to the DMDC to initiate an electronic fingerprint capture policy in 1997.

In an initial pilot program, DMDC saved an estimated $8 million with 25,000 military retirees living in overseas locations. The program confirmed DMDC's suspicion that military benefits were still being collected on deceased retirees when many failed to appear to enroll their fingerprint in the new identification system (Dunn, 1998). Since 1998, the DMDC has been capturing the right index fingerprint of all active-duty, reserve, and retired military personnel as well as survivors receiving a military annuity. This potential enrollment pool is some 3 million people. The fingerprint is captured during the routine issuance (or reissuance) of military identification cards at some 900 DMDC sites. DMDC stores electronic copies of these fingerprints in a comprehensive database known as the Defense Enrollment Eligibility Reporting System (DEERS). DMDC does not store any copies of fingerprints on the actual military identification card. DEERS can be accessed if a person's identity needs to be authenticated.

U.S. Naval Criminal Investigative Service (NCIS)

NCIS has been testing a fingerprint-based system to provide secure data and voice communications with undercover agents who are unable to risk physical meetings. The small pilot began in November 1999 and concluded in February 2000. Participation in the program has been voluntary. As such, no privacy or consent issues arose with the five enrollees. Further, as NCIS agents, all have security clearances, all current and potential enrollees already have a great deal of personal information, including fingerprints, on record and are thus less likely to oppose participating in such a system.

The NCIS system consists of a number of laptops, each with an externally adapted scanner. Fingerprint data are stored in the laptop itself for verification. That is, the system does not include a central server that keeps a database of fingerprint data. The laptops' basic input/output system (BIOS) has been modified, so the computers will not operate without verification by the agent to whom the laptop has been assigned or by the system administrator.

According to one technician, during the trial these systems had only one technical problem, which involved the laptop, not the biometric element. He noted, however, that the scanners are sensitive to weather and lighting. Specifically, it is difficult to get a fingerprint reading in direct sunlight.

AN NCIS special agent involved with the test expressed his satisfaction with the system and his hope that its use would continue and expand beyond the trial period. He noted that the system has "proven itself to be reliable and do what we wanted it to do, which is to protect the information and the communications."

Face Recognition for Countersurveillance

DoD is currently working with a commercial vendor on a special project known as "Face Recognition for Countersurveillance and One-to-Many Identification of Antigovernment Factions." This project, many aspects of which are classified, has been fielded at select military installations overseas. The face recognition system is not a facial verification system but rather strives to identify the individual by performing a "one-to-many" search. In operational terms, the face recognition system takes the facial input of a subject, generates a template, compares this template to a database and then provides a list of potential candidates as an output.

Established in the mid-1990s, the vendor gained its first contract through the National Institute of Justice. Its first project assisted law enforcement officials in tracking gang members. The military saw the potential for this technology to support intelligence collection by helping to track terrorists and insurgents overseas. In 1995, the Air Force became the first military service to work with the vendor on a face recognition system. Since then, DoD has worked with the vendor on several face recognition projects.

COMMERCIAL BIOMETRIC PROGRAMS

Riverside Health System Employees Credit Union, Newport News, Virginia

The Riverside Health System includes three hospitals, 10 nursing homes, five wellness and fitness centers, three retirement communities, and 210 doctors' offices. The credit union serving this system is small, employing three staff members. In 1997, the credit union felt that the government was moving away from "dead-tree technology" and toward all-electronic transactions. As part of the process, the credit union decided to add an electronic kiosk known as the "Money Buddy"—in effect, a 24-hour automated branch. The Money Buddy allows customers to print statements of their accounts, transfer funds between accounts, apply for loans, make loan payments, and print checks for withdrawals. Money Buddy requires account numbers and fingerprints for customer access. No card is necessary. The system was in place by July 1998. Money Buddy acts as "force multiplier" for the credit union

Credit union customers include many military families: Langley AFB, Norfolk Naval Station, U.S. Coast Guard Reserve Training Center (Yorktown), the Army's Fort Eustis, and NASA's Langley Research Center are all in the Riverside Health System area. Perhaps because of the clientele's familiarity and comfort with on-base security measures, most have come to see the fingerprint system as more protective than invasive. Privacy issues have been insignificant, and both customers and management are very comfortable with the system. Account holders' fingerprint data are deleted when they close their accounts.

The system has encountered very few problems. Of 536 customers currently scanned in, about 10 are unable to use the system and use PINs instead (no card is required). Two of the 10 are a plumber and a carpenter (both military veterans) who have worn their fingers almost completely smooth. The other eight have fingerprints with horizontal lines, which apparently cause problems at the resolution level used by the fingerprint scanner. Smaller problems include climatic and occupational factors that alter individuals' fingerprints. Specifically, skin dryness that accompanies health care workers' frequent hand washings can lead to fingerprint distortions. These dryness problems can be resolved by adding a bit of oil to the finger by rubbing it behind the ear.

General Services Administration (GSA) and Citibank

Since May 1999, the U.S. General Services Administration (GSA) has been using fingerprint verifications for computer workstation security as part of a nine-month pilot study. The system being tested requires no passwords or PINs and currently has 500 enrollees. Few problems have cropped up with the system, and those that have appeared are consistent with the problems found elsewhere—cuts, rings, etc., that can distort fingerprint images, whether at the time of the initial reference scan or during subsequent scans.

Some GSA employees, through their union, initially raised some concerns about privacy, but these subsided following an explanation of the system, its benefits, and the safeguards in place for employee data. Specifically, the fingerprint templates collected are encrypted when stored in the GSA database and on the associated chip card.

Visa, San Francisco, California

Visa has been exploring the use of biometrics for the past 15 years, starting with dynamic signature recognition. To date, Visa has run trials and pilot programs with most forms of biometrics, including finger, voice, hand, iris, face, and signature, in its search for what it considers the best biometrics approach to be used with their services.

Visa's operations in San Francisco use a hand geometry recognition system for their internal physical access. A program official interviewed reported there have been no failures with the hand geometry component as part of the physical security system at the Visa headquarters building. Hand geometry readers limit access to certain restricted locations within the facility.

Visa has delayed its push to use biometrics with its credit card operations for several reasons. First, Visa has been waiting for the price of biometrics systems to fall before the company pursues them in earnest. Second, Visa has yet to find the right vendor and biometrics approach. Although Visa believes a finger scan to be the proper approach, they have not been as impressed with some of the results from trials with several vendors. Third, Visa realizes that a standard approach is necessary among credit card services, so that each does not use different vendors and different readers, which would make it difficult for business customers to implement the system.

Visa also sees a need to continue educating the public about biometrics. In its trials and market research, Visa found very little stigma associated with the use of biometrics. Visa still needs to educate the public about data protection. Another aspect of education is to inform the public, including potential criminal elements, that using a dismembered hand or finger for unauthorized access will fail.

Kroger Supermarkets, Texas

Kroger, a national supermarket chain, has recently completed a year-long trial of fingerprint biometrics recognition in conjunction with check cashing in six of its stores. Because of the trial program's success, Kroger fielded the system in 250 stores nationwide by the first quarter of 2000.

Kroger uses a fingerprint scanning system. In each store, approximately 45 percent of all Kroger customers write checks to pay for their purchases. These customers are given the opportunity to participate in the fingerprint-scanning program. A Kroger executive said that about 8,000–10,000 customers per store participated in the test phase of the program. Kroger believes it will have similar numbers as it expands the biometric program to all of its stores. Kroger reportedly has been pleased with the program's performance as well as the overall reduction of check fraud in its stores. In the six trial stores, approximately 1,000 incidents of check fraud took place each month before Kroger implemented the system. The pilot stores have had zero incidents of check fraud since implementation. This dramatic drop in the incidence of fraud has created a large enough savings that the system should pay for itself within a year.

A Kroger executive explained that Kroger experienced very little negative reaction from customers to the use of fingerprint scanning. Customers have been pleased at the hassle-free process of paying by check. No longer do they need to show an ID card but simply put their finger on the scanner and within a second the process is over.

After a customer places his finger on the scanner, the data collected are matched to a local database on the store workstation containing the records of that store's customers. If the fingerprint is not in the store's local database, the computer searches the main Kroger database off site. If the customer is found in the main database, the individual is identified and the local database then receives the customer's record for future transactions.

Kroger was initially concerned that their senior citizen customer base might express concerns over the program but found that this group was the most enthusiastic about the biometrics system. The seniors are highly motivated with regard to fraud and security and welcomed the fingerprint scanners. Kroger did discover that seniors tend to have drier hands than younger people, and that at first hampered getting a good read on the scanner. Kroger made adjustments and now the system operates well for this customer group.

PHYSICAL ACCESS

Columbia Presbyterian Hospital, New York, N.Y.

The hospital has used a TimeLink hand geometry scanner since 1997 to monitor time and attendance and control physical access. Time and attendance measurement is the primary use of the system, which was made necessary by perceived inaccuracies in bookkeeping. In the first year of the system's operation, fraud reduction led to an estimated savings of more than $1 million. The payroll department and TimeLink maintain the system, which currently has 8,000 enrollees.

A few minor problems have been associated with the hand geometry system. Employees expressed some initial privacy concerns (e.g., "Is this taking my fingerprints?"), but the use of the system has become routine at this point, and these concerns appear to have subsided. Scanning problems can occur when the lens or scanning surface become dirty. The hospital's housekeeping department is responsible for keeping the scanning units clean. The wearing of bandages, long sleeves, and other add-ons can also distort the hand image, whether during the initial scan or during subsequent access scans. Employees' data are deleted from the system immediately with their separation from the hospital.

Universal Air Cargo Security Access System, Chicago, Illinois

The Universal Air Cargo Security Access System is a pilot program at O'Hare International Airport, the world's busiest airport. The security system is sponsored by the Federal Aviation Administration (FAA), the Chicago Department of Aviation, the American Trucking Association, and 25 trucking companies and 22 airlines. SecurCom is the systems integrator. The system features fingerprint scanners by Identix, smart cards by Schlumberger, and database software by Oracle.

Knowing that approximately 60 percent of all air cargo going through O'Hare is transported on passenger flights, airport officials and concerned parties realized this represented a potentially large security loophole through which terrorists could plant explosives or other contraband. With Universal system, truck drivers who already have a security clearance are given a biometrically encoded smart card, which contains data regarding the contents of their trucks and the number of the back door's seal. A cleared inspector encodes the card by biometrically signing off on the cargo. On arrival at the airport, both the driver and the seal are verified and biometrically accepted by the cargo attendant. The truck's payload is accepted into the airport for further processing following this verification.

The first phase of Universal's pilot, in which 12 airlines and 52 trucking companies with about 500 drivers participated, was completed in March 1999. The second phase, which will bring Newark International Airport on line as well, began in February 2000.

Other projects at O'Hare include the installation of fingerprint/smart card readers for access to the U.S. Customs area in the international terminal. SecurCom is replacing access readers at O'Hare with similar readers. Midway Airport is next in line. Currently, O'Hare's 50,000 badged employees are using a magnetic card/PIN system.

University of Georgia, Athens, Georgia

The University of Georgia has one of the longest-running biometric applications in the United States. The university started using biometrics as an identifier in 1972. Since then, it has continued to employ various biometric strategies for identification purposes.

They now use hand recognition systems for physical access purposes.

The University of Georgia saw a need to restrict access to its student dining facilities. Prior to 1972, the university used a punch card system that was ineffective and easily circumvented. In 1972, they implemented a biometric hand reader. Although problems persisted with students being able to fake the process by moving their hands while the system measured their fingers, it was an improvement over the punch cards. In 1995, they implemented a three-dimensional biometric hand geometry system. At that time, 5,400 students—those on the university's meal program—were added to the new system. Because of the success of the program, both in terms of student reaction and of curtailing unauthorized access to the dining facilities, the university decided to expand the system to address other physical access needs. The same hand geometry system is used to grant access to the 17 residence halls for 5,600 students (most of these students are in the meal program). Since 1998, the University of Georgia has required all 31,000 students to enroll in the hand geometry program, which prevents unauthorized access to the university's sport and recreation facilities.

The implementers of the system at the university were surprised at the relatively few complaints from the students. The university leaders introduced the system to the students as something "state of the art" and that they as a school were "pushing the future." They found that students were pleased to be taking part in something unique.

A university official mentioned that they do have problems with some aspects of the hand biometric system. If individuals have extremely small hands or have had broken hands, it can render the system unusable. The official also explained that for a successful reading to take place, the individual must be comfortable with the system. At the University of Georgia, many students use the system multiple times every day, so they become quickly accustomed to the procedure and have few problems.

Good Shepherd Hospital, Barrington, Illinois

In 1995, for approximately six months, the hospital used a voice recognition system to control access to the operating rooms. The system provided so many false negatives (and irate surgeons) that it was disconnected and the magnetic card system it replaced was reinstalled. It had been thought that voice recognition would be an easy way to control access to the operating rooms without forcing surgeons to carry a card or key.

SOCIAL SERVICES

At least eight states use large-scale biometric applications in social service programs: Arizona, California, Connecticut, Illinois, Massachusetts, New Jersey, New York, and Texas.[2] We reviewed programs in Connecticut, Texas, and California. These states were chosen because of the availability of information. They are in no way a representative sampling of the states' activities. States beginning implementation of a biometric identification program include Florida, North Carolina, and Pennsylvania, while 18 more states are pursuing legislation regarding such matters.

Los Angeles Country AFIRM Program

The Los Angeles County Department of Public Social Services (DPSS) program targeted participants in the Aid to Families with Dependent Children (AFDC) and Food Stamp programs. The biometrics program is known as Automated Fingerprint Image Reporting and Match (AFIRM). It was designed to prevent fraud through duplicate participation or "double-dipping," defined as the same person enrolled in a system multiple times using multiple aliases. AFIRM uses fingerprint matching provided by PrinTrak, and EDS handles system management.

DPSS used ink-and-paper fingerprints as early as 1986. In 1988, a steering committee approved automated fingerprint matching. By 1991, DPSS launched a pilot program using automated fingerprinting. By the end of 1994, the program had been launched at all 25 DPSS district offices. The program includes 300,000 people who must be fingerprinted. These include adults receiving AFDC pay-

---

[2]For more information on biometrics programs operated by state social services departments, an excellent starting point is Connecticut State DSS (2000).

ments, minor parents receiving payments, and adults collecting payments for children. The biometric consists of templates of two index fingers. The data are not shared with law enforcement officials under any circumstances.

Prior to launching the program, DPSS staff explained the process to their clients and educated them as to what the system would entail. DPSS made appointments for enrollment. Those unable to make their appointments were given an additional 10 days. After that, if an adult failed to report for his appointment, adult benefits were cut off, although children's benefits continued. If an adult continued to refuse to enroll, the case was referred to the fraud units.

According to a DPSS review, most participants did not feel inconvenienced by the biometric. Rather they believed the biometric would be effective in reducing fraud, which most felt was a positive step. Of 137 cases sampled for noncompliance, 76 percent were deemed fraudulent. The DPSS felt that the biometric program saved a substantial amount of money, about $66 million in savings over 26 months.

Texas Department of Human Services (DHS)

In 1995, the Texas state legislature mandated implementation of electronic imaging as part of Texas's initiative to reduce fraud in public assistance programs. Based on TDHS's research of available electronic imaging systems, fingerprint imaging was determined to be the most reliable and affordable technology for identification verification purposes.

Texas's finger imaging program, the Lone Star Image System, was developed to deter duplicate participation in the Food Stamp and Temporary Assistance for Needy Families (TANF) programs. A pilot project of the Lone Star Image System began in October 1996 in 10 offices in the San Antonio region, enrolling more than 85,000 clients.

Based on the success of the San Antonio pilot program, federal approval for full statewide implementation was given in May 1998 and implementation was completed in August 1999.

Adults (over 18 years of age) and minor heads of household receiving Food Stamps or TANF are required to provide finger images when they apply or recertify. Fingerprint imaging of two index fingers and a digital photograph of the individual constitute the enrollment record. More than 400 Lone Star Image System enrollment stations can be found throughout the state, including some mobile stations at temporary offices. Finger image enrollments are routinely purged after six months of inactivity.

Although the system has not caught many individuals committing fraud, TDHS estimates that the system saves $6.36 million each year by deterring potential duplicate recipients. TDHS estimates that the incidence of duplicate participation in the Food Stamp program is about one half of one percent of the total caseload.

In 1997, the Texas legislature instructed TDHS to plan a pilot project allowing clients to provide finger images instead of a PIN at the point of sale when accessing benefits under the Lone Star Card/electronic benefit transfer (EBT) program. This program employs a debit card instead of Food Stamp coupons or paper checks in distributing Food Stamp and TANF benefits. However, in 1999, the Texas legislature did not approve funding for the pilot project, stating that the technology of biometrics at point of sale was not sufficiently mainstream at the time.

Results of a survey conducted by TDHS showed that 89 percent of program participants thought biometrics were a good idea and 81 percent think using finger images instead of PINs at point of sale is a good idea.

Lack of standardization among the biometric venders is a major problem. For example, DHS has worked with Kroger supermarkets to develop a finger imaging at point-of-sale joint pilot project. However, interoperability problems—merging TDHS's system with Kroger's existing finger imaging check authorization program would require Kroger to use a separate fingerprint scanner—added cost and complexity to the process.

Connecticut Department of Social Services (DSS)

Legislation drafted in 1995 funded the study and eventual deployment of a fingerprint biometric to prevent welfare fraud. Connecticut wanted to create a system that would deter dual enrollments, including enrollments in neighboring states. It is not uncommon for the same person to illegally participate in several states' entitlement programs at the same time through the use of aliases and forged identification documents. Accordingly, DSS selected a biometric with an eye toward compatibility with the neighboring states of New York and New Jersey. As it turned out, the states' templates are not compatible, making interstate comparisons somewhat complicated.

Connecticut has 24,000 general-assistance enrollees and 60,000 AFDC clients in its system. The program was implemented in 16 regional offices, 20 town general-assistance offices, and the DSS Hartford office. It uses centralized image storage and retrieval. The cards can be used in one-to-one verification or one-to-many identification using the network of databases.

In addition to the fingerprints stored, each card and file carried a photograph and signature of the recipient that can be manually matched by social services staff to verify the recipient.

Prior to implementing its program, the DSS conducted an extensive education campaign. Despite these efforts, some members of the legislature vigorously opposed the program. In addition, since its establishment, the state has received three refusals to participate made on religious grounds. These cases were resolved by an administrative decision to allow the three persons to use alternative identification means. DSS conducted a survey of program participants and found that the majority approve of the Connecticut biometric program. More than 80 percent of those responding stated that they favored the program.

Connecticut's vision for biometrics includes using the ID card in EBT transactions, point-of-sale devices for disbursement of medical services, and distribution of Food Stamp benefits through food retailers.

DSS estimated its first year operating costs at $2.6 million with an estimated savings (from deterrence) of $7.5 million.

Illinois Department of Human Services

In 1994, the Illinois legislature approved a study of the use of biometric scanning to detect and deter fraud in programs administered by the Illinois Department of Public Aid. Officials tested retinal scanning in two offices downstate and fingerprint scanning in three Chicago offices. Fingerprints were required in the test offices for cash disbursements but not for Food Stamps or medical payments. The department was very satisfied with the fingerprint system and dissatisfied with the retina scanning system.

In July 1997, the department was partially incorporated into the Illinois Department of Human Services (DHS). This organizational change led to changes in information technology personnel. Currently, the system is partially operational, and DHS is pursuing a decision to expand electronic fingerprinting statewide. A full-time staff member has been hired to provide technical support for the system.[3]

Social Services Summary

All three programs have had to deal with privacy concerns, and each had a handful of objections raised on privacy grounds. Each of the databases was explicitly declared to be inaccessible by law enforcement officials. All use secure designs to protect against hackers and have procedures in place to prevent unauthorized disclosure of information. Connecticut's DSS believes that answers to privacy concerns can be found in the careful packaging of the implementation legislation, use of the biometric only for the social services program integrity, and a secure design of the biometric system to protect from unauthorized disclosure.

Use of biometrics in the social services sector will continue to expand. States are seeking to make their systems more robust, both in terms of interstate compatibility and with additional applications.

IMMIGRATION AND LAW ENFORCEMENT

U.S. Immigration and Naturalization Service (INS)

INS deployed its Immigration and Naturalization Service Passenger Accelerated Service System (INSPASS) in 1993. INSPASS is based on hand geometry (but it was also designed to allow the use of finger-

---

[3]See also Illinois State DHS (1997).

prints as an alternative). The prototype installations were at JFK, Newark, and Pearson (Toronto) International Airports. Additional deployments include Miami, Los Angeles, San Francisco, Dulles, Vancouver, B.C., and other high-volume international airports.

More than 85,000 people are currently enrolled in this frequent international traveler program, and more than 200,000 transactions have been processed since its installation. INS, in cooperation with the Department of State, determines the rules for who may participate in INSPASS. Citizens of the United States, Canada, Bermuda, legal permanent residents of the United States, most landed immigrants in Canada, and Visa Waiver Pilot Program countries with visa classifications B-1, D-1, TN, WB, and some nonimmigrants in classes A, E, G, and L who travel to the United States on business three or more times a year or who are diplomats, representatives of international organizations, or airline crews from pilot program nations may voluntarily enroll in the INSPASS Program. Access to INSPASS is not available to anyone with a criminal record or to aliens who require a waiver of inadmissibility to enter the United States.

As of last year, approximately 35,000 American and foreign users have voluntarily registered in the system. Los Angeles International Airport alone enrolls 40–50 new users per day, and 40–100 users take advantage of the LAX INSPASS kiosk each day. Roughly 25,000 international passengers go through INS control at LAX daily. Gaining border access with INSPASS typically takes less than one minute, while waiting in line for manual passport stamping can take up to 45 minutes for U.S. citizens and two hours for foreigners.

Travelers who register false reads are sent to see an INS inspector and can be locked out of the system for four hours. There is a problem with people who have small hands (e.g., Japanese flight attendants have been particularly problematic at LAX). People who have no right hands use their left hands upside-down.

Sarasota County Detention Center, Sarasota, Florida

Since 1998, the Sarasota County Detention Center has used an iris recognition system, created by IriScan, to verify the identities of its approximately 750 inmates. As inmates are brought into the detention center, a device scans their irises, and they are enrolled in the system. Currently, the detention center uses the system only to verify its inmates when they enter and when they leave.

Within the detention center, the inmates also use photo ID cards for internal verification. In the past, inmates would steal the cards of those inmates about to be released in an attempt to assume their identity and escape. Since the IriScan system was implemented, eight inmates have been caught by the iris recognition system while trying to escape using stolen identities. Another individual was falsely arrested and released when the iris scan revealed he was not the person the police wanted—the wanted suspect was a recently released inmate with his iris template still on file.

The Sarasota County Detention Center is pleased with the system. The whole system, including implementation, cost around $6,000 and has already proved its value through the foiled escapes. The system also reduces the need to have forensic experts assist in proving the identity of an individual by reading fingerprints. The iris recognition system takes less than a second to verify the individual and provides positive identification at any time of the day.

The database allows for input into a comment section where data concerning warrants can be maintained, which helped the detention center identify an individual who had three additional outstanding warrants.

DoD DNA Specimen Repository for Remains Identification

The DNA Repository for Remains Identification along with the Armed Forces DNA Identification Laboratory make up the DoD DNA Registry.[4] The DNA Registry, a Division of the Office of the Armed Forces Medical Examiner, helps the military identify remains of soldiers killed in combat or missing in action. High-velocity weapons and the lethality of the modern battlefield often destroy any chances of using fingerprints or dental records. DNA, however, can almost always be used to identify remains. Although most times the armed forces can identify the dead based on various records, DNA identifi-

[4]For an excellent discussion of the DoD DNA Registry, see Weedn (1998). Dr. Weedn was the founder and for seven years the program manager of the DoD DNA Registry.

cation provides closure for the family and the biological proof of death required by life insurance companies.

This issue came to a head as the military prepared for Operation Desert Storm with the potential for large numbers of casualties. The Dover AFB, Delaware, mortuary facilities were expanded, but medical officials were concerned about the ability to identify the dead. DNA techniques had been pursued by the military in its efforts to identify servicemembers missing in action from Vietnam, Korea, and even World War II, but this was a slow process that required the military to find close relatives and obtain samples from them in an attempt to match them to DNA samples from the deceased.

Army pathologists were convinced that the need to identify large numbers of dead service personnel had to be addressed and that a military DNA registry could provide a suitable solution. The Army leadership also became convinced of the utility of such a registry and lobbied for it. In December 1991, authorization and appropriations for the DNA program were received. Since June 1992, DoD has required all military inductees and all active-duty and reserve personnel to provide DNA samples for its DNA Repository at the time of enlistment, reenlistment, annual physical, or preparation for operational deployment. The DNA Repository also contains samples from civilians and foreign nationals who work with the U.S. military in arenas of conflict. DoD stores the samples in freezers at the DNA Specimen Repository in Gaithersburg, Maryland.

Implementation of the program in the Army, which is the executive agent of the DNA program for DoD, was not without controversy. Everyone was concerned about privacy, from DoD officials to policymakers to the media. The program office began to hold meetings to educate military personnel about the purpose of the program and the privacy protections that would be used to ensure that DNA data would not be otherwise employed. The education campaign worked, and at all levels military personnel have participated in the program. To date, those who have refused to participate in the DNA registry have been forced to leave the service.[5] As of 1998, only three

---

[5]On March 17, 1997, a DoD directive permitted the armed service branches to exempt certain members from the mandatory DNA collection requirement to accommodate religious practices.

servicemembers have refused to submit samples, as opposed to some 1.3 million servicemembers who have complied.[6]

Undoubtedly, a number of servicemembers are unwilling participants but have chosen to trust the Army rather than leave the service. In addition to the education campaign, other announcements had to be made about the program. In particular, on June 14, 1995, DoD placed "system of records" notice in the *Federal Register* announcing the establishment of this new system containing personal information (Weedn, 1998, p. 354). This announcement, required by the Privacy Act, needed to be approved by DoD's Privacy Board. It was, after deliberations that took 18 months.

Another major issue for the program was how long to keep the DNA records. One might assume that they would just be pulled when a servicemember leaves the military, but apparently similarity of servicemembers' names or SSNs as well as clerical error raised the risk of pulling the wrong record. It is also time-consuming to search the repository continually for individual records, particularly when the records number more than 3 million.

Originally, DoD's policy called for destruction of DNA records after 75 years. However, in 1996, DoD changed the destruction schedule to 50 years, to be compatible with standards for military health records.[7] This 50-year period ensures that no servicemember remains in the armed forces when his DNA record is pulled from the database. Also, in 1996, DoD amended its policy to permit servicemembers to request that their DNA samples be destroyed when they leave the service. In other words, servicemembers can opt out of the database. Once a servicemember makes such a request, DoD has six months to destroy the DNA records.

DoD's strict policy on sharing of the specimens ensures that DNA specimens can only be used for

- remains identification

---

[6]See Weedn (1998, p. 354) See also *Mayfield v. Dalton*, 901 F.Supp. (D. Hawaii 1995) (dismissing all the claims of two Marines who refused to participate in DNA program on grounds that it infringed on their constitutionally protected privacy rights).

[7]Apparently, the time period was changed as a "technical correction" (Weedn, 1998, p. 351).

- internal quality-control purposes
- consensual uses, and
- other limited uses as compelled by law.

This last category includes a court order authorizing sharing investigation or the approval of the DoD General Counsel or Assistant Secretary of Defense for Health Affairs for prosecution of a felony. The specimens cannot be used without consent for any other purpose, such as paternity suits or genetic testing. In addition, the specimens are considered confidential medical information and are covered by federal laws and military regulations on privacy. This policy has been tested by numerous federal agencies who have asked for access to the data, primarily for law enforcement purposes.

SOCIAL SECURITY

The Use and Misuse of Social Security Numbers

The controversial history of Social Security numbers (SSNs) provides an important case study on the subject of citizens' privacy rights vis-à-vis federal, state, and local government. When first devised in 1935, the SSN was issued to workers exclusively for Social Security Administration (SSA) accounting purposes. The cards as originally issued noted, "Not for Identification Purposes." By 1943, however, an Executive Order required that "all Federal components use the SSN 'exclusively' whenever the component found it advisable to set up a new identification system for individuals." (U.S. Social Security Administration, 1998.)[8] Since then, the SSN has been at the center of a public debate about whether there should be a U.S. national identification card.

In 1999, the U.S. General Accounting Office (GAO) submitted a report detailing government and commercial use of SSNs to a House subcommittee. At that time, Congress was considering legislation regulating the use of SSNs in response to public concerns about organizational use of SSNs and the role of the SSN in the growing phenomenon of identity theft (GAO, 1999, p. 1). The GAO report

---

[8]Information on the history of the SSN is available at http://www.ssa.gov/history.

found that "no single federal law regulates the overall use of SSNs." Rather, a number of laws require the use of SSNs for specific applications (e.g., Medicaid, Food Stamps, commercial driver's licensing programs), while other laws restrict the SSNs' use. Significantly, the GAO found that "no federal law . . . imposes broad restrictions on businesses' and state and local governments' use of SSNs when that use is unrelated to a specific federal requirement." (GAO, 1999, p. 2.)[9]

The use of the SSN has attracted increasing legislative attention as high-speed data processing systems have made SSN use more commonplace in both the public and private sectors. Organizations and agencies that GAO consulted cited the usefulness of the SSN as an identifier that transcends state boundaries and name changes and is easily used for transferring data among bodies (*e.g.*, credit bureaus to banks, HMOs to hospitals). These organizations and agencies made their belief clear to the GAO that "their entities would be negatively affected if federal laws were enacted restricting use of SSNs" (GAO, 1999, p. 12).

It is this very ease of data transfer, however, that has led some members of Congress and various watchdog groups to support legislation restricting the use of SSNs and making identity theft a crime.[10] Depending on which side one believes, the information revolution heralds either a new era of convenience, ease of transaction, and security or the advent of a governmental-industrial Big Brother with far more knowledge about U.S. citizens than is warranted.

---

[9] For examples of people and institutions opposed to the widespread use of SSNs or other forms of national identification, see Moore (1997) and SCAN (2000). See Miller and Moore (1995) and Garfinkel (2000, pp. 16–35) for a discussion of the SSN and function creep.

[10] See, e.g., Identity Theft and Assumption Deterrence Act of 1998 (P.L. 105-318).

Appendix C

LEGAL ASSESSMENT: LEGAL CONCERNS RAISED BY THE U.S. ARMY'S USE OF BIOMETRICS[1]

EXECUTIVE SUMMARY

From the legal perspective, the Army's use of biometrics raises concerns in three critical areas. These include statutory and administrative law concerns, constitutional law concerns, and international law concerns. The major statutory and administrative law structure that applies to Army use of biometrics is imposed by the Privacy Act of 1974, which regulates the collection, maintenance, use, and dissemination of personal information by federal government agencies. Accordingly, this Act is examined in great detail.

Army use of biometrics implicates constitutional rights involving informational privacy and physical privacy under the Bill of Rights as well as religious freedom under the First Amendment. To help the Army understand these rights, background information is presented and important cases dealing with these issues are discussed. *Whalen v. Roe*, the Supreme Court's leading case on informational privacy, is analyzed from the perspective of what the Army can learn from this case. To help the Army understand the real-world setting in which these rights operate in the context of biometric applications, this appendix includes discussion of two legal challenges raised on reli-

---

[1]The principal author of this Appendix, John D. Woodward, Jr., Esq., appreciates the helpful comments and insights provided by Stewart A. Baker, Esq., Robert R. Belair, Esq., Arthur S. Di Dio, M.D., J.D., Professor Steve Goldberg of the Georgetown University Law Center, Kristina Larson, Catherine A. Szilagyi, Esq., and Shirley C. Woodward, Esq. Their assistance in no way implies their endorsement of the views presented in this appendix or acquiescence in any mistakes contained herein.

gious grounds to state-mandated biometric applications in New York and Connecticut. To help the Army understand how other federal agencies view possible legal objections to biometric applications, recent experience of the Federal Bureau of Investigation (FBI) in this area is detailed.

As the Army increasingly operates in the international arena, Army use of biometrics could raise issues of international law. To help the Army understand how a major international law related to privacy can affect biometric applications, the possible impact of the European Union Data Protection Directive on U.S. Army biometric applications in European Union member states is assessed. Similarly, the possible impact of laws of other foreign nations is also addressed.

After surveying the legal landscape related to biometrics, this review concludes that Army use of biometric applications in the United States should not encounter any significant legal obstacles, provided the Army complies with the mandates of the Privacy Act and the teachings of the Supreme Court. To ensure this compliance, Army leadership can call on the many institutional assets within DoD and the Army who are experienced and skilled in dealing with privacy issues. These assets can do a case-by-case analysis of the biometric application and determine exactly what needs to be done legally to ensure compliance. In sum, while biometrics is a new technology, the Army has an existing framework that can accommodate legal requirements.

As Army biometric applications venture overseas, the Army leadership must consider international law issues raised by these applications on a case-by-case basis. The impact of the European Union Data Protection Directive on the U.S. Army as a data collector in EU member states is not entirely clear. The United States and the EU agreed in July 2001 on a "safe harbor" framework, which provides U.S. organizations a means of satisfying the directive's requirement that personal data is afforded an "adequate" level of privacy protection. This safe harbor framework is designed primarily for private sector entities, however, and it does not appear that the Army would currently be eligible to join the safe harbor. It appears likely, however, that the Army's use of biometrics will comply with the directive by virtue of falling within one of its exceptions, although continued attention is required because the various exceptions and exemptions to compliance have yet to be definitively interpreted. Although the EU directive is a new and controversial privacy law, the Army has a framework in place to monitor the issues raised by the directive and to provide the Army with the necessary legal support to ensure compliance.

STATUTORY AND ADMINISTRATIVE LAW CONCERNS

The Privacy Act of 1974

Overview. The Privacy Act of 1974 regulates the collection, maintenance, use, and dissemination of personal information by federal government agencies, including DoD and the U.S. Army.[2] It serves as the basis for both the DoD Privacy Program and Army Privacy Program.[3] The Act requires the Office of Management and Budget (OMB) to prescribe guidelines and regulations for federal agencies to use in implementing the Privacy Act and provide continuing assistance for and oversight of the implementation of the Privacy Act by agencies.[4]

In broad terms, the Privacy Act gives certain rights to the "data subject"—or the individual who provides personal information—and places certain responsibilities on the "data collector"—the agency collecting the personal information. The Privacy Act balances a federal agency's need to collect, use, and disseminate information about individuals with the privacy rights of those individuals. In particular, the Act tries to protect the individual from unwarranted invasions of privacy stemming from a federal agency's collection, maintenance, use, and dissemination of personal information about the individual.[5]

---

[2]The Privacy Act of 1974, codified at 5 U.S.C. § 552a, as amended, went into effect on September 27, 1975. See Department of Justice (1998 and 1999).

[3]The DoD Privacy Program is issued under the authority of DoD Directive 5400.11, dated June 9, 1982. DoD 5400.11-R, dated August 31, 1982, establishes regulations for the implementation of the DoD Privacy Program. AR 340-21, dated July 5, 1985, establishes regulations for the implementation of the Army Privacy Program.

[4]5 U.S.C. § 552a(v)(1) and (2).

[5]There are several things the Privacy Act does not do. For example, the Privacy Act does not regulate the collection, maintenance, use, and dissemination of personal information by state and local government agencies. See *Ortez v. Washington County,*

Along these same lines, the DoD Privacy Program "is intended to provide a comprehensive framework regulating how and when the Department collects, maintains, uses, or disseminates personal information on individuals. The purpose of the Program is to balance the information requirements and needs of the Department against the privacy interests and concerns of the individual" (DoD, 2000c). Similarly, the Army Privacy Program sets out "the privacy rights of individuals and the Army's responsibilities for compliance with operational requirements established by the Privacy Act."[6]

The Privacy Act's basic provisions, reflected in both the DoD Privacy Program and the Army Privacy Program,[7] include

- restricting federal agencies from disclosing personally identifiable records maintained by the agencies;
- requiring federal agencies to maintain records with accuracy and diligence;
- granting individuals increased rights to access records about them maintained by federal agencies and to amend their records, provided they show that the records are not accurate, relevant, timely, or complete; and
- requiring federal agencies to establish administrative, technical, and policy safeguards to protect record security.[8]

As these basic provisions suggest, the Privacy Act sets forth a so-called "code of fair information practices" requiring federal agencies, as data collectors, to adopt minimum standards for collection, use, maintenance, and dissemination of records. It also requires that

---

Oregon, 88 F.3d 804, 811 (9th Cir. 1996). The Privacy Act does not regulate personal information held by private sector entities. See 5 U.S.C. § 552a; 5 U.S.C. § 552f (definition of "agency"). See also *Gilbreath v. Guadalupe Hosp. Found.*, 5 F.3d 785, 791 (5th Cir. 1993). The Privacy Act does not apply when the individual, or data subject, is not a U.S. citizen or an alien lawfully admitted for permanent residence. See 5 U.S.C. § 552a(a)(4).

[6]AR 340-21 at ¶ 1-5.

[7]Unless otherwise indicated, the Privacy Act provisions discussed in this RAND Report apply to DoD and the U.S. Army.

[8]See, e.g., Cate (1997, p. 77) and Department of Justice (1998) at "Individual's Right of Access," "Individual's Right of Amendment," and "Agency Requirements."

agencies publish detailed descriptions of these standards and the procedures used to implement them. Data collector responsibilities are discussed below.

Applicability to Biometrics. Although the Privacy Act does not specifically mention "biometrics," our analysis strongly suggests that the Act can include Army biometric applications. As the Act applies to a "record" that is "contained in a system of records," the threshold issue to resolve is whether biometric identification information, whether in the form of an image file or a template file, falls within the Act's broad definition of record. The Act defines "record" as:

> [A]ny item, collection, or grouping of information about an individual that is maintained by an agency, including, but not limited to, his education, financial transactions, medical history, and criminal or employment history and that contains his name, or the identifying number, symbol, or *other identifying particular assigned to the individual, such as a finger or voice print or a photograph. . . .*[9]

The OMB *Guidelines* explain that "record" means "any item of information about an individual that includes an individual identifier" and "can include as little as one descriptive item about an individual."[10] The Court of Appeals for the Third Circuit has affirmed the *Guidelines*' definition, finding that "record" includes "any information about an individual that is linked to that individual through an identifying particular."[11] The Court of Appeals for the District of Columbia has stressed that the Privacy Act only protects "information that actually describes the individual in some way."[12]

As explained in the main body of this report, biometrics are distinctive individual identifiers. They are "identifying" and they are "particular" to an individual. Moreover, fingerprint and voiceprint, two of the examples cited in the Act's definition of "record," are physical characteristics. As such, they fall within the definition of

---

[9]See 5 U.S.C. § 552a(a)(4) (emphasis added).

[10]See OMB (1987) (quotations omitted). See also Department of Justice (1998) at "Definitions: D. Record."

[11]*Quinn v. Stone*, 978 F.2d 126, 133 (3d Cir. 1992).

[12]*Tobey v. N.L.R.B.*, 40 F.3d 469, 471-73 (D.C. Cir. 1994).

biometrics. Accordingly, biometrics satisfy the Privacy Act's definition of "record."

To fall within the Privacy Act, the record must be "contained in a system of records." The Act defines "system of record" as:

> [A] group of any records under the control of any agency from which information is retrieved by the name of the individual or by some identifying number, symbol, or other identifying particular assigned to the individual. . . .[13]

OMB's *Guidelines* explain that a system of records exists when two conditions are met. First, there must be an "indexing or retrieval capability using identifying particulars [that is] built into the system." Second, the agency must "in fact, retrieve records about individuals by reference to some personal identifier" (OMB, 1987, and Department of Justice, 1998, at "Definitions: E. System of Records"). Commenting on these OMB *Guidelines*, the Court of Appeals for the District of Columbia has explained that a federal agency must not only have "the capability to retrieve information indexed under a person's name, but the agency must in fact retrieve records in this way in order for a system of records to exist."[14]

To determine if an Army biometric application is a record contained in a system of records, the Army must do a case-by-case analysis of each application examining how the biometric is used. For some applications, it is possible that the Privacy Act would not be implicated because the record is not contained in a system of records. For example, the Army's Fort Sill pilot program did not implicate the Privacy Act because, while the biometrically protected digital cash card provided to Army basic trainees was arguably a record, the fingerprint template was stored only on the card. It was not contained in any system of records, such as a central database. On the other hand, some applications will implicate the Act. Such an application would include biometric identification information combined with information about an individual that can be retrieved by an identifying particular, like a biometric.

---

[13]See 5 U.S.C. § 552a(a)(5).

[14]*Henke v. United States Dep't of Commerce*, 83 F.3d 1453, 1460 n.12 (D.C. Cir. 1996).

In cases where an Army biometric application implicates the Privacy Act, the Army must make certain that it complies fully with the Act's provisions. In ensuring this compliance, the Army can draw on many existing institutional assets who have extensive experience in Privacy Act matters. These assets include the Defense Privacy Board,[15] the Assistant Secretary of Defense (Comptroller), the Defense Privacy Office, the DoD General Counsel, the Army Assistant Chief of Staff for Information Management, the Army General Counsel, the Army Judge Advocate General, the Army Office of the Deputy Chief of Staff for Personnel, OMB, and many others.

The Privacy Act's major requirements are explained below.

The "No Disclosure Without Consent Rule." The Privacy Act prohibits a federal agency from "disclos[ing] any record which is contained in a system of records by any means of communication to any person, or to another agency, except pursuant to a written request by, or with the prior written consent of, the individual to whom the record pertains . . . [subject to certain exceptions discussed below]."[16] This provision is known as the "No Disclosure Without Consent Rule."

While the "No Disclosure Without Consent Rule" applies, the Act contains twelve enumerated exceptions to this rule.[17] The exceptions to the "No Disclosure Without Consent" Rule are as follows:

(1) The "Intra-Agency Need to Know" Exception

(2) The "Required Freedom of Information Act (FOIA) Disclosure" Exception

(3) The "Routine Use" Exception

(4) The "Bureau of the Census" Exception

---

[15]Membership of the Defense Privacy Board consists of the Director of the Defense Privacy Office, who sits as Executive Secretary, and Representatives designated by the Secretaries of the Military Departments, the Assistant Secretary of Defense (Comptroller) (whose designee serves as Chairperson), the Assistant Secretary of Defense (Manpower, Reserve Affairs, and Logistics), the DoD General Counsel, and the Director of the Defense Logistics Agency. See DoD 5400.11-R, ¶ 6.1.

[16]See 5 U.S.C. § 552a(b).

[17]See 5 U.S.C. § 552a(b)(1)-(12).

(5) The "Statistical Research" Exception

(6) The "National Archives" Exception

(7) The "Law Enforcement Request" Exception

(8) The "Individual Health or Safety" Exception

(9) The "Congressional" Exception

(10) The "General Accounting Office" Exception

(11) The "Judicial" Exception

(12) The "Debt Collection Act" Exception.

These broadly structured exceptions are discussed below.

The "Intra-Agency Need to Know" exception. This applies when officers and employees of the federal agency maintaining the record have a need for the record in the performance of their duties.[18] In the case of medical records, the Army construes this exception somewhat narrowly by restricting what is disclosed. For example, the applicable Army regulation provides that when "medical information is officially requested for a use other than patient care, only enough information will be provided to satisfy the request."[19]

The "Required Freedom of Information Act ("FOIA") Disclosure" exception. This exception provides that the Privacy Act cannot be used to prohibit a disclosure that the FOIA requires.[20]

The "Routine Use" exception. As for disclosure of a record, a "routine use" means "the use of such record for a purpose which is compati-

---

[18]See 5 U.S.C. § 552a(b)(1). See, e.g., *Britt v. Naval Investigative Serv.*, 886 F.2d 544, 549 n.2 (3d Cir. 1989) (approving, as "intra-agency need to know" exception, disclosure of investigative report to Britt's Marine Corps Reserve commanding officer "since the Reserves might need to reevaluate Britt's access to sensitive information or the level of responsibility he was accorded"); *Beller v. Middendorf*, 632 F.2d 788, 798 n.6 (9th Cir. 1980) (approving disclosure of record revealing servicemember's homosexuality by Naval Investigative Service to commanding officer for purpose of reporting "a ground for discharging someone under his command").

[19]AR 40-66, ¶ 2.2(e), dated May 3, 1999.

[20]See 5 U.S.C. § 552a(b)(2). See also *Greentree v. United States Customs Serv.*, 674 F.2d 74, 79 (D.C. Cir. 1982) (Privacy Act is not to "be used as a barrier to FOIA access").

ble with the purpose for which it was collected."[21] The Privacy Act requires that the federal agency publish in the *Federal Register* "each routine use of the records contained in the system, including the categories of users and the purpose of such use."[22] Thus, the federal government agency must satisfy two requirements for a proper routine use disclosure: The routine use must be "compatible" and constructive notice must be given by publication of the agency's routine use in the *Federal Register*.[23]

According to OMB, compatibility encompasses functionally equivalent uses and other uses that are necessary and proper.[24] The federal judiciary has not settled on a uniform interpretation of compatibility. For example, the Court of Appeals for the District of Columbia has adopted a broadly construed "common usage" requiring only that "a proposed disclosure would not actually frustrate the purposes for which the information was gathered."[25] On the other hand, the Court of Appeals for the Third Circuit put forth a narrower construction: a "concrete relationship or similarity, some meaningful degree of convergence, between the disclosing agency's purpose in gathering the information and its disclosure."[26] In cases where the federal judiciary must determine the legality of a federal agency's routine use, the judiciary gives deference to the federal government agency's construction of its routine use.[27]

---

[21]See 5 U.S.C. §§ 552a(b)(3); 552a(a)(7) (definition of "routine use").

[22]See 5 U.S.C. § 552a(e)(4)(D).

[23]Some federal courts have determined that a third requirement exists: Actual notice of the routine use must be given to the individual at the time the information is collected from him. See *United States Postal Service v. National Ass'n of Letter Carriers*, 9 F.3d 138, 146 (D.C. Cir. 1993) (stating that "[a]lthough the statute itself does not provide, in so many terms, that an agency's failure to provide employees with actual notice of its routine uses would prevent a disclosure from qualifying as a 'routine use,' that conclusion seems implicit in the structure and purpose of the Act"); *Covert v. Harrington*, 876 F.2d 751, 754-56 (9th Cir. 1989).

[24]See OMB (1987), 52 Fed. Reg. 12,990, 12,993.

[25]*United States Postal Service v. National Ass'n of Letter Carriers*, 9 F.3d 138, 144 (D.C. Cir. 1993).

[26]*Britt v. Naval Investigative Service*, 886 F.2d 544, 555 (3d Cir. 1989).

[27]See, e.g., *Department of the Air Force, Scott Air Force Base, Ill. v. FLRA*, 104 F.3d 1396, 1402 (D.C. Cir. 1997); *FLRA v. U.S. Dep't of Treasury*, 884 F.2d 1446, 1451 (D.C. Cir. 1989).

Two important types of "compatible" routine uses frequently occur with respect to law enforcement. First, in the context of investigations and prosecutions, law enforcement agencies routinely share law enforcement records with each other.[28] Second, agencies may routinely disclose any records indicating a possible violation of law, regardless of the purpose for collection, to law enforcement agencies for purposes of investigation and prosecution.[29] For example, the Army has published a so-called "law enforcement blanket routine use" which applies to every record system maintained within the Army, unless a specific exception is made. One such exception is that the "law enforcement blanket routine use" does not apply to the "Armed Forces Repository of Specimen Samples for the Identification of Remains" system of records, which includes "specimen collections from which a DNA typing can be obtained."[30]

The "law enforcement blanket routine use" provides that:

> In the event that a system of records maintained by [the Army] to carry out its functions indicates a violation or potential violation of law, whether civil, criminal or regulatory in nature, and whether arising by general statute or by regulation, rule, or order issued pursuant thereto, the relevant records in the system of records may be referred, as a routine use, to the appropriate agency, whether Federal, state, local, or foreign, charged with the responsibility of investigating or prosecuting such violation or charged with enforcing or implementing the statute, rule, regulation, or order issued pursuant thereto.[31]

---

[28]See, e.g., OMB (1987, 40 Fed. Reg. 28,955) (proper routine use is "transfer by a law enforcement agency of protective intelligence information to the Secret Service"); see also 28 U.S.C. § 534 (authorizing Attorney General to exchange criminal records with "authorized officials of the Federal Government, the States, cities, and penal and other institutions").

[29]See OMB (1987, 40 Fed. Reg. 28,953); see also 28 U.S.C. § 535(b) (1994) (requiring agencies of the Executive Branch to expeditiously report "[a]ny information, allegation, or complaint" relating to crimes involving government officers and employees to United States Attorney General).

[30]See 63 Fed. Reg. 10,205, March 2, 1998. See also Armed Forces (2000). See also Appendix B, Program Reports, DoD DNA Specimen Repository for Remains Identification.

[31]*Preamble to the Department of Army Privacy Act Systems of Records Notice,* available at http://www.defenselink.mil/privacy/notices/army/army_preamble.html. Additional Army blanket routine uses are published at this site.

Because of its "potential breadth," the routine use exception is a controversial provision of the Privacy Act.[32] For example, it has been called "a huge loophole"[33] that has been used by federal agencies to justify almost any use of the data (Cate, 1997, p. 78, footnote omitted). The two law enforcement routine use exceptions discussed above have been criticized on the ground that they circumvent the more restrictive requirements of the routine use exception.[34]

Moreover, Congress can always mandate additional new "routine uses" for agencies, which the affected agencies must establish as "routine uses" (OMB, 1987, 40 Fed. Reg. 28,954). For example, Congress has mandated the establishment of a federal "Parent Locator Service" within the Department of Health and Human Services and requires federal agencies to comply with requests from the Secretary of HHS for addresses and places of employment of absent parents.[35]

The "Bureau of the Census" exception. This exception is for disclosure of information made to the U.S. Bureau of the Census for purposes of planning or carrying out a census or related activity pursuant to statute.[36]

The "Statistical Research" exception. This exception permits disclosure of information to entities that will use the information for statistical research or a reporting record. The information must be transferred to the entity in a form that is not individually identifiable.[37]

---

[32]See Department of Justice (1998), "Conditions of Disclosure to Third Parties: B. Twelve Exceptions to the 'No Disclosure Without Consent' Rule: 3. 5 U.S.C. § 552a(b)3 (routine uses)."

[33]See Cate (1997, p. 78), citing David Flaherty, the former British Columbia Data Protection Commissioner (footnote omitted).

[34]See Department of Justice (1998) (citing *Privacy Commission Report* at 517–518; Britt, 886 F.2d at 548 n.1 (dictum); *Covert*, 667 F. Supp. at 739, 742 (dictum)). See also Privacy International, *Privacy and Human Rights 1999* (asserting that the Privacy Act's effectiveness is "significantly weakened by administrative interpretations [of the routine use exception]").

[35]See 42 U.S.C. § 653.

[36]See 5 U.S.C. § 552a(b)(4).

[37]See 5 U.S.C. § 552a(b)(5).

The "National Archives" exception. This limited exception permits disclosure of records that have sufficient historical or other value to warrant consideration for their preservation by the U.S. government.[38]

The "Law Enforcement" exception. This exception provides for disclosure of information to federal law enforcement agencies and allows an agency, "upon receipt of a written request, [to] disclose a record to another agency or unit of State or local government for a civil or criminal law enforcement activity."[39]

The "Individual Health or Safety" exception. This exception permits disclosure of information pursuant to a showing of compelling circumstances affecting the health or safety of an individual.[40] For example, dental records on several individuals could be released to identify an individual injured in an accident.

The "Congressional" exception. This exception applies to disclosure of information to the House of Representatives and the Senate or, to the extent of matter within its jurisdiction, any committee or subcommittee thereof, any joint committee of Congress or subcommittee of any such joint committee.[41]

The "General Accounting Office" exception. This exception applies to disclosure of information to the Comptroller General in the course of the performance of the duties of the General Accounting Office.[42]

The "Judicial" exception. This exception applies to court orders requiring disclosure.[43] It prevents the Privacy Act from "be[ing] used to block the normal course of court proceedings, including court-ordered discovery."[44] Some disagreement exists as to what exactly constitutes a "court order." The issue centers on whether a subpoena issued by a court clerk, as opposed to the court itself,

[38]See 5 U.S.C. § 552a(b)(6).

[39]See OMB (1987, 40 Fed. Reg. 28,948, 28,955); 5 U.S.C. § 552a(b)(7).

[40]See 5 U.S.C. § 552a(b)(8).

[41]See 5 U.S.C. § 552a(b)(9).

[42]See 5 U.S.C. § 552a(b)(10).

[43]See 5 U.S.C. § 552a(b)(10).

[44]See *Clavir v. United States*, 84 F.R.D. 612, 614 (S.D.N.Y. 1979).

should qualify under this exception. A Defense Privacy Board Advisory Opinion has concluded that, "[a] subpoena signed by a clerk of a Federal or State court, without specific approval of the court itself, does not comprise an 'order of a court of competent jurisdiction' for purposes of nonconsensual disclosures [under the judicial exception]. . . . [D]isclosure of records [under this exception] requires that the court specifically order disclosure" (DoD, 2000b). Similarly, the Court of Appeals for the District of Columbia has held that a subpoena routinely issued by a court clerk—such as a federal grand jury subpoena—is not a "court order" within the meaning of this exception because it is not "specifically approved" by a judge.[45]

The "Debt Collection Act" exception. The Debt Collection Act of 1982 authorized this disclosure exception. It permits agencies to disclose bad debt information to credit bureaus. Before disclosing this information, however, agencies must complete a series of due process steps designed to validate the debt and to offer the individual an opportunity to repay it.[46]

Under the Privacy Act, rights are personal to the individual who is the subject of the federal agency record. These rights cannot be asserted by others on behalf of the aggrieved individual.[47]

Agency Responsibilities. *Overview:* The Privacy Act places certain responsibilities on the data collector. These responsibilities include publishing information about the systems of records in the data collector's charge, giving notice to data subjects of the uses to which the data will be put, and safeguarding data.

Publication: Among the responsibilities the Privacy Act places on the data collector, it requires an "agency that maintains a system of records" to "publish in the *Federal Register* upon establishment or revision a notice of the existence and character of the system of

---

[45]See *Doe v. DiGenova*, 779 F.2d 74, 77-85 (D.C. Cir. 1985).

[46]See 5 U.S.C. § 552a(b)(11); OMB (1987, 48 Fed. Reg. 15,556-60).

[47]See, e.g., *Parks v. IRS*, 618 F.2d 677, 684-85 (10th Cir. 1980) (which holds that a union lacks standing to litigate its members' Privacy Act claims); *Word v. United States*, 604 F.2d 1127, 1129 (8th Cir. 1979) (which holds that a criminal defendant lacks standing to allege Privacy Act violations regarding use at trial of medical records concerning third party); *Dresser Indus. v. United States*, 596 F.2d 1231, 1238 (5th Cir. 1991) (which holds that a company lacks standing to litigate employees' Privacy Act claims).

records."[48] This notice, which is known as a "Privacy Act Systems of Records Notice," must include

- the name and location of the system;
- the categories of individuals about whom records are maintained in the system;
- the categories of records maintained in the system;
- each routine use of the records contained in the system, including the categories of users and the purpose of such use;
- the policies and practices of the agency regarding storage, retrievability, access controls, retention, and disposal of the records;
- the title and business address of the agency official responsible for the system of records;
- the agency procedures whereby an individual can be notified at his request if the system of records contains a record pertaining to him;
- the agency procedures whereby an individual can be notified at his request how he can gain access to any record pertaining to him contained in the system of records, and how he can contest its content; and
- the categories of sources of records in the system.[49]

The Army has 249 systems of records for which such notice must be published (DoD, 2000a). These range from "Official Personnel Folders and General Personnel Files" (AAFES 0401.04) to "Individual Health" (AAFES 0405.11) to "Carpool Information/Registration System" (A0001SAIS) and many others (DoD, 2000a).

The Privacy Act permits a federal agency to promulgate rules to exempt systems of records from certain parts of the Privacy Act if certain conditions are met. One such condition is if the system of records is maintained as a principal function by a law enforcement

---

[48]See 5 U.S.C. § 552a(e)(4).

[49]See 5 U.S.C. § 552a(e)(4)(A)-(I).

agency and the records were compiled for law enforcement purposes.[50] Other conditions include if the system of records contains classified information;[51] investigatory material compiled for law enforcement purposes;[52] material maintained and used solely as statistical records;[53] investigatory material compiled solely for the purpose of determining suitability, eligibility, or qualifications for federal civilian employment, military service federal contracts or access to classified information;[54] and other conditions.[55]

As Army use of biometrics will likely lead to the establishment of new systems of records and revisions to old systems, the Army must comply with this Privacy Act Systems of Records Notice requirement. As its 249 systems of records suggest, the Army has ample experience in doing so.

Notice: The Privacy Act requires the data collector to give notice[56] to the data subject informing him of four factors:

- The authority that authorizes the solicitation of the information and whether disclosure of such information is mandatory or voluntary.
- The principal purpose or purposes for which the information is intended to be used.
- The routine uses that may be made of the information.
- The effects on the data subject if any, of not providing all or any part of the requested information.[57]

---

[50] 5 U.S.C. § 552a(j)(2).

[51] 5 U.S.C. § 552a(k)(1).

[52] 5 U.S.C. § 552a(k)(2).

[53] 5 U.S.C. § 552a(k)(4).

[54] 5 U.S.C. § 552a(k)(5).

[55] See, e.g., 5 U.S.C. § 552a(k)(4) ("U.S. Secret Service" exception); 5 U.S.C. § 552a(k)(6) ("testing materials" exception).

[56] This notice may be given (1) on the actual form which the data collector uses to collect the information desired or (2) on a separate form that can be retained by the individual. See 5 U.S.C. § 552a(e)(3).

[57] See 5 U.S.C. § 552a(e)(3)(A)-(D). The authority may be granted by statute or executive order of the President. See 5 U.S.C. § 552a(e)(3)(A).

In its biometric applications, the Army will likely comply with the Privacy Act's notice requirement during the biometric enrollment process, when it first collects the biometric identification information from the data subject. As an institution that collects much information from many individuals, the Army has extensive experience in satisfying the notice requirement.

Data Safeguarding: The Privacy Act requires the data collector to "establish appropriate administrative, technical, and physical safeguards to insure the security and confidentiality of records." Similarly, the Act requires the data collector "to protect against any anticipated threats or hazards to their security or integrity which could result in substantial harm, embarrassment, inconvenience, or unfairness to any individual about whom information is maintained."[58]

As this provision of the Act makes clear, the data collector must put in place appropriate safeguards to protect information in its databases. However, as a federal district court has explained, "[t]he Privacy Act does not make administrative agencies guarantors of the integrity and security of materials which they generate."[59] Instead, "the agencies are to decide for themselves how to manage their record security problems, within the broad parameters set out by the Act."[60] Accordingly, the data collectors "have broad discretion to choose among alternative methods of securing their records commensurate with their needs, objectives, procedures, and resources."[61]

The Senate Report accompanying the Privacy Act supports this judicial view:

> The Committee recognizes the variety of technical security needs of the many different agency systems and files containing personal information as well as the cost and range of possible technological

---

[58]See 5 U.S.C. § 552a(e)(10).

[59]*Kostyu v. United States*, 742 F.Supp. 413, 417 (E.D. Mich. 1990) (which holds that alleged lapses in IRS security resulting in disclosure of information to public were not willful and intentional as required to establish Privacy Act violation).

[60]*Id.*

[61]*Id.*

> methods of meeting those needs. The Committee, therefore, has not required [ ] in this Act a general set of technical standards for security of systems. Rather, the agency is merely required to establish those administrative and technical safeguards which it determines appropriate and finds technologically feasible for the adequate protection of the confidentiality of the particular information it keeps against purloining, unauthorized access, and political pressures to yield the information to persons with no formal need for it.[62]

The Senate Report stressed that data collectors have flexibility in deciding appropriate safeguards:

> The [Privacy] Act . . . provides reasonable leeway for agency allotment of resources to implement this subsection. At the agency level, it allows for a certain amount of "risk management" whereby administrators weigh the importance and likelihood of the threats against the availability of security measures and consideration of cost.[63]

While a breach of database security and confidentiality can be harmful or embarrassing to the data collector, both the agency and the employee responsible for the breach can be found legally liable for a Privacy Act violation. This legal liability can include civil liability for the agency and criminal liability for an agency official. Civil liability for such a breach attaches when "the agency has acted in a manner which was intentional or willful."[64] The federal judiciary has interpreted this phrase "to require a showing of fault 'somewhat greater than gross negligence.' "[65]

Similarly, criminal liability, in the form of a misdemeanor, attaches for such a breach when an "officer or employee of an agency, who by virtue of his employment or official position, has possession of, or

---

[62]*Id.* (citing S.Rep. No. 93-1183, reprinted in 1974 U.S. Code Cong. & Admin. News 6916, 6969).

[63]*Id.* (citing S.Rep. No. 93-1183, reprinted in 1974 U.S. Code Cong. & Admin. News 6916, 6969).

[64]See 5 U.S.C. § 552a(g)(4); *Pilon v. United States Department of Justice*, 796 F.Supp. 7, 12 (D.D.C. 1992); *Kostyu*, 742 F.Supp. at 416.

[65]*Kostyu*, 742 F.Supp. at 416.

access to, agency records [covered by the Privacy Act], and who knowing that disclosure of the specific material is so prohibited, willfully discloses the material in any manner to any person or agency not entitled to receive it."[66] Likewise, criminal liability can attach when an "officer or employee of any agency [ ] willfully maintains a system of records without meeting the notice requirements of [the Privacy Act]."[67]

In implementing its biometric applications that fall under the Privacy Act, the Army must meet all of the Act's many requirements. Successfully meeting these requirements will require a comprehensive, coordinated effort drawing on DoD's Privacy Act institutional assets as well as the Army's appropriate technical, security, law enforcement, and administrative assets. However, the Army has complied with the Privacy Act in the past, is complying now, and should continue to comply in the future. Fortunately, the Army has a seasoned and experienced structure already in place to ensure Privacy Act compliance.

Additional Safeguards. *The Computer Matching and Privacy Act of 1988*: The Computer Matching and Privacy Act of 1988 ("The Computer Matching Act") amended the Privacy Act by adding new provisions regulating federal agencies' computer matching practices and placing requirements on the agencies.[68] A computer match is done by using a computer program to search an agency's files for infor-

[66]See 5 U.S.C. § 552a(i)(1). Certain exemptions apply. See, e.g., 5 U.S.C. § 552a(j).

[67]See 5 U.S.C. § 552a(i)(2). Certain exemptions apply. See, e.g., 5 U.S.C. § 552a(j).

[68]See 5 U.S.C. § 552a(o) (1988). See also Turkington and Allen (1999, pp. 362–363), from which this section is largely drawn. The Computer Matching and Privacy Protection Act of 1988 amended the Privacy Act to add several new provisions. These include 5 U.S.C. § 552a(a)(8)-(13), (e)(12), (o), (p), (q), (r), and (u). These provisions add procedural requirements for agencies to follow when engaging in computer matching activities; provide matching subjects with opportunities to receive notice and to refute adverse information before having a benefit denied or terminated; and require that agencies engaged in matching activities establish Data Integrity Boards to oversee those activities. These provisions became effective on December 31, 1989. OMB's guidelines on computer matching should be consulted in this area. See 54 Fed. Reg. 25,818-29 (1989). Subsequently, Congress enacted the Computer Matching and Privacy Protection Amendments of 1990, which further clarify the due process provisions found in subsection (p). OMB's proposed guidelines on these amendments appear at 56 Fed. Reg. 18,599-601 (proposed April 23, 1991). See Department of Justice (1998) at "Computer Matching."

mation associated with or indexed by a personal identifier, such as a name or SSN. The information thus obtained can then be compared with information in the databases of another federal agency. In this way, discrepancies and inconsistencies might be discovered that point to fraud in government benefits, for example.

DoD participates in approximately 25 computer matching programs with various different government agencies (DoD, 2000d). For example, DoD has a "Debt Collection" matching program in effect with the Department of Education. The purpose of this program is to identify and locate federal personnel who are delinquent on payments to certain programs administered by the Department of Education.

For all of its matching programs, DoD must meet the Computer Matching Act's requirements, which basically involve entering into formal agreements with the exchanging agencies,[69] verifying independently the accuracy of data received before any official action is taken,[70] providing notice in the *Federal Register* prior to conducting or revising a computer matching program,[71] and establishing a Data Integrity Board to monitor implementation and compliance with the Act.[72]

Because a personal identifier in the form of a biometric could implicate the Computer Matching Act, the Army will need to study the Act closely to determine whether the Army's specific biometric application is implicated. As with the Privacy Act, the Army can call on many existing institutional assets with experience in matters pertaining to the Computer Matching Act.

Administrative Regulation: From the administrative regulatory perspective, Congress can follow two well-worn policy paths when dealing with a public policy issue involving a new technology, such as biometrics. It can take the direct route and pass legislation regulating Army use of the technology or it can delegate its authority to

---

[69]See 5 U.S.C. § 552a(o)(1)(A-D).

[70]See 5 U.S.C. § 552a(o)(1)(E).

[71]See 5 U.S.C. § 552a(o)(1)(D).

[72]See 5 U.S.C. § 552a(u)(1).

the appropriate administrative agencies within DoD. The delegation route is the road most frequently traveled. However, even though the Army, specifically, and DoD, in general, are well-equipped with expertise, experience, and institutional memory, they still face enormous challenges in designing, formulating, and implementing government policy for biometric applications. In addition, numerous competing groups (many well-organized and some politically influential) will want to press their claims in this public policy process.[73]

The Army should bear in mind that Congress, through the legislative process, can require the Army to satisfy additional conditions related to its biometric applications. For example, Congress could go beyond the Privacy Act and place additional prohibitions on disclosure of biometric identification information and further restrict sharing.

The Uniform Code of Military Justice (UCMJ) and Privacy Protections. Historically, the Supreme Court has long recognized that differences between the civilian and military criminal law systems exist. The Court has stated that "[m]ilitary law, like state law, is a jurisprudence which exists separate and apart from the law which governs in our federal establishment."[74] Most important, the Court has acknowledged that the military criminal law system, embodied by the UCMJ, can impose restrictions on a servicemember's rights. However, the UCMJ does not strip a servicemember of his or her constitutional rights. As the Court of Military Appeals has observed: "[I]t is apparent that the protections in the Bill of Rights, except those which are expressly or by necessary implication inapplicable, are available to members of our armed forces."[75] For example, the Supreme Court has explained that the special demands of "military life do not, of course, render nugatory in the military context the guarantees of the First Amendment."[76] In the context of Army biometric applications, however, the UCMJ does not seem to provide

---

[73]See generally Goldberg (1994).

[74]*Burns v. Wilson*, 346 U.S. 137, 140 (1953).

[75]*United States v. Jacoby*, 29 C.M.R. 244 (C.M.A. 1960) (*citing Burns v. Wilson*, 346 U.S. 137 (1953); *Shapiro v. United States*, 69 F.Supp. 205 (Ct. Cl. 1947); *United States v. Hiatt*, 141 F.2d 664 (3d Cir. 1944).

[76]*Goldman v. Weinberger*, 475 U.S. 503 (1986).

servicemembers with any greater privacy rights beyond what is in the U.S. Constitution.

CONSTITUTIONAL LAW CONCERNS

Introduction

Beyond the specific individual rights provided by statutory and regulatory regimes, the Constitution, through its Bill of Rights, protects individual privacy rights. These constitutionally protected privacy rights consist of physical, decisional, and informational privacy rights. These privacy rights do not pose a constitutional barrier to Army biometric applications, provided that the Army follows the guidance of the Supreme Court as explained in this section.

Overview. As the Army expands its biometric applications and requires more and more members of the Army community to provide biometric identification information, it is likely that someone required to participate will refuse and (1) face disciplinary action within the military justice system, if the refuser is under its jurisdiction, and/or (2) file a federal lawsuit, claiming his constitutional rights are violated. The military is no stranger to such litigation.[77] This section of the appendix begins by examining how the federal judiciary views the military. It then explores the major bases of any legal challenges that might be brought on constitutional grounds.

Judicial Deference to Military. Several Supreme Court decisions have established that the federal judiciary views the nation's military as uniquely different from civilian society. For example, then-Justice William H. Rehnquist explained that the Supreme Court "ha[s] repeatedly held that 'the military is, by necessity, a specialized society separate from civilian society.' "[78] In the preparation and performance of its duties, "the military must insist upon a respect for duty

[77]*Id.* (involving Air Force officer who brought lawsuit against the Secretary of Defense claiming that the uniform regulation that prevented him from wearing his yarmulke infringed on his constitutional rights).

[78]*Id.* at 507 (citing *Parker v. Levy,* 417 U.S. 733, 743 (1974); *Chappell v. Wallace,* 462 U.S. 296, 300 (1983); *Schlesinger v. Councilman,* 420 U.S. 738, 757 (1975); *Orloff v. Willoughby,* 345 U.S. 83, 94 (1953)). See also *Burns v. Wilson,* 346 U.S. at 140.

and a discipline without counterpart in civilian life."[79] "[W]ithin the military community, there is simply not the same [individual] autonomy as there is in the larger civilian community."[80]

The Supreme Court recognizes that "the military authorities have been charged by the Executive and Legislative Branches with carrying out our Nation's military policy."[81] Moreover, the Supreme Court has observed that the "courts [are] 'ill-equipped to determine the impact upon discipline that any particular intrusion upon military authority might have.' "[82] Therefore, "[j]udicial deference . . . is at its apogee when legislative action under the congressional authority to raise and support armies and make rules and regulations for their governance is challenged."[83] The Court has determined that "[j]udges are not given the task of running the Army."[84] Rather, "[t]he responsibility for setting up channels through which . . . grievances can be considered and fairly settled rests upon the Congress and upon the President of the United States and his subordinates. The military constitutes a specialized community governed by a separate discipline from that of the civilian."[85] Because the military is so different from the civilian community, "[o]rderly government requires that the judiciary be as scrupulous not to interfere with legitimate Army matters as the Army must be scrupulous not to intervene in judicial matters."[86]

This judicial deference that the federal judiciary gives to the military suggests that the federal courts may be somewhat reluctant to intrude into proper Army concerns related to biometrics. However, the federal courts will not hesitate to protect the constitutional rights of an individual. To help ensure that it receives this deference, the

---

[79]*Schlesinger v. Councilman*, 420 U.S. at 757. See also *Brown v. Glines*, 444 U.S. 348, 354 (1980).

[80]*Goldman v. Weinberger*, 475 U.S. at 507 (citing *Parker v. Levy*, 417 U.S. at 751).

[81]*Id.*

[82]*Id.* (citing *Chappell v. Wallace*, 462 U.S. at 305, quoting Warren, 1962).

[83]*Goldman v. Weinberger*, 475 U.S. at 508 (quoting *Rostker v. Goldberg*, 453 U.S. 57, 70 (1981)).

[84]*Orloff v. Willoughby*, 345 U.S. at 93-94.

[85]*Id.*

[86]*Id.*

Army should be prepared to demonstrate that each of its biometric applications serves a worthwhile and useful military purpose.

What Privacy Rights Does the U.S. Constitution Recognize?

Survey of Privacy Scholarship. Jurists and scholars have long grappled with defining what privacy is and explaining what privacy should be (Cate, 1997, pp. 19–31).[87] In 1879, Judge Thomas M. Cooley, in his classic treatise on torts, included "the right to be let alone" as a class of tort rights, asserting that "[t]he right to one's person may be said to be a right of complete immunity" (Hixson, 1987, p. 30, and Goldberg, 1994, p. 114). Echoing and popularizing Cooley's phrase, Samuel D. Warren and Louis D. Brandeis (1890), in their landmark law review article, *The Right to Privacy,* voiced their view of privacy as a "right to be let alone." Brandeis, as a Supreme Court Justice, used this phrase in his famous dissent in *Olmstead v. United States,* declaring that the Founding Fathers "conferred, as against the Government, the right to be let alone—the most comprehensive of rights and the right most valued by civilized men."[88] Privacy as the "right to be let alone" has a positive appeal and commendable simplicity, but the phrase has been criticized in that "legally, it offers no guidance at all. Coveting an indefinable right is one thing; enforcing it in a court of law is another" (Alderman and Kennedy, 1995, p. xiv).

More recent scholarship also offers insight into privacy. For example, Ruth Gavison (1980, pp. 421, 428) offers what is perhaps the extreme privacy model: "[P]rivacy is a limitation of others' access to an individual . . . [I]n perfect privacy no one has any information about X, no one pays any attention to X, and no one has physical access to X." Privacy includes a control aspect—"control we have over information about ourselves" (Fried, 1970, p. 140), "control over who can sense us" (Parker, 1974, pp. 275, 281, internal quotation marks omitted), or "control over the intimacies of personal identity" (Gerety, 1977, pp. 233, 236) Based on her survey of the extensive privacy literature, Professor Lillian R. Bevier (1995, pp. 455, 458, foot-

---

[87]While a detailed discussion of the many facets of privacy is beyond the scope of this report, an excellent starting point is Turkington and Allen (1999).

[88]277 U.S. 438, 478 (1928) (Brandeis, J., dissenting); see also Cate (1997, p. 57) and Goldberg (1994, p. 114).

note omitted) concluded, "[p]rivacy is a chameleon-like word, used denotatively to designate a range of wildly disparate interests—from confidentiality of personal information to reproductive autonomy—and connotatively to generate goodwill on behalf of whatever interest is being asserted in its name."[89]

Constitutional Background. The word "privacy," like the word "biometrics," is nowhere to be found in the text of the U.S. Constitution. However, without making explicit reference to privacy, the Constitution nonetheless protects certain privacy interests.[90] The Bill of Rights contains these protections in the First Amendment rights of freedom of speech, press, religion and association; the Third Amendment prohibition against the quartering of soldiers in one's home; the Fourth Amendment right to be free from unreasonable searches and seizures; the Fifth Amendment right against self-incrimination; the Ninth Amendment's provision that "[t]he enumeration in the Constitution, of certain rights, shall not be construed to deny or disparage others retained by the people"; and the Tenth Amendment's provision that "[t]he powers not delegated to the United States by the Constitution, nor prohibited by it to the States, are reserved to the States respectively, or to the people."

What then is the constitutional right to privacy and how does it affect biometrics used in U.S. Army applications? The answer to the first part of the question is legally fuzzy. As a federal appellate court has recently observed, "[w]hile the Supreme Court has expressed uncertainty regarding the precise bounds of the constitutional 'zone of privacy,' its existence is firmly established."[91]

Most modern constitutional privacy interests have their roots in the Due Process Clause of the Fourteenth Amendment. This clause says that no state shall "deprive any person of life, liberty, or property, without due process of law." For more than 100 years, these words have been interpreted by the Supreme Court to contain a substantive

---

[89]See also Murphy (1996, p. 2381).

[90]Or "zones of privacy," to use Justice Douglas's term. *Griswold v. Connecticut,* 381 U.S. 479, 484 (1965) (holding unconstitutional a state statute that criminalized the sale of contraceptives to married couples).

[91]*In re Crawford,* 1999 U.S. App. LEXIS 24941, *7 (9th Cir. 1999) (citing *Whalen v. Roe,* 429 U.S. 589, 599-600 (1977); *Griswold v. Connecticut,* 381 U.S. at 483).

protection that "bar[s] certain government actions regardless of the fairness of the procedures used to implement them."[92]

Three Forms of Privacy. The Supreme Court has stressed that "there is a realm of personal liberty which the government may not enter."[93] This realm, or zone of privacy, consists of rights that are "fundamental" or "implicit in the concept of ordered liberty"[94] or as a later Court would put it, "deeply rooted in this Nation's history and tradition."[95] In what specific areas of the zone of privacy is the government forbidden entry? In considering privacy interests, the Court has implicitly categorized privacy as taking three distinct forms (Allen, 1991, p. 175).[96] These three forms of privacy include:

- Physical privacy or freedom from contact with other people or monitoring agents. Physical privacy enjoys its greatest constitutional protection under the Fourth Amendment freedom from unreasonable search and seizure.
- Decisional privacy or the freedom of the individual to make private choices about the personal and intimate matters that affect him without undue government interference. The Court has found that the individual is constitutionally protected in "personal decisions relating to marriage, procreation, contraception, family relationships, child rearing, and education."[97]

---

[92]*Planned Parenthood of Southeastern Pennsylvania v. Casey*, 505 U.S. 833, 846 (1992) (quoting *Daniels v. Williams*, 474 U.S. 327, 331 (1986)).

[93]*Id.* at 847.

[94]*Griswold v. Connecticut*, 381 U.S. at 500 (Harlan, J., concurring) (quoting *Palko v. Connecticut*, 302 U.S. 319, 325 (1937)).

[95]*Moore v. City of East Cleveland*, 431 U.S. 494, 503 (1977). These terms have been criticized for lack of clarity. See, e.g., Bork (1990, p. 118) "[T]he judge-created phrases specify no particular freedom, but merely assure us, in sonorous phrases, that they, the judges, will know what freedoms are required when the time comes."

[96]At least one scholar has more broadly categorized the Supreme Court's interpretation of constitutional protections for individual privacy as falling into four areas—"expression and association, searches and seizures, fundamental decisionmaking, and informational privacy" (Cate, 1997, p. 52).

[97]*Planned Parenthood v. Casey*, 505 U.S. at 851. In determining the commonality of these personal decisions and why they deserve constitutional protection, the Court, through Justice Sandra Day O'Connor's opinion in *Casey*, explained that:

> These matters, involving the most intimate and personal choices a person may make in a lifetime, choices central to personal dignity

- Informational privacy or freedom of the individual to limit access to certain personal information about oneself. The Court of Appeals for the Ninth Circuit has defined this phrase as "the individual interest in avoiding disclosure of personal matters."[98] Privacy scholar Alan Westin defines it as "the claim of individuals . . . to determine for themselves when, how, and to what extent information about them is communicated to others" (Westin, 1967, p. 337, citing Scott and Jarnagin, 1868, pp. 457–507). Similarly, Professor Lawrence Lessig (1999, p. 143), drawing heavily on the scholarship of Ethan Katsh (1995, p. 228), has defined privacy in this context as "the power to control what others can come to know about you." As Lessig explains, others can acquire information about you by monitoring and searching. Monitoring refers to that part of one's daily existence that others see, observe, and can respond to. Searching refers to that part of one's life that leaves or is a record that can later be scrutinized (Lessig, 1999, p. 143). Noting both quantity and quality aspects to informational privacy, a federal appellate court has phrased it in terms of, "control over knowledge about oneself. But it is not simply control over the quantity of information abroad; there are modulations in the quality of knowledge as well."[99]

The Army's use of biometrics could potentially require its soldiers, civilian employees, independent contractors, along with many other

---

> and autonomy, are central to the liberty protected by the Fourteenth Amendment. At the heart of liberty is the right to define one's own concept of existence, of meaning, of the universe, and of the mystery of human life. Beliefs about these matters could not define the attributes of personhood were they formed under compulsion of the State.

[98]*Doe v. Attorney General*, 941 F.2d 780, 795 (9th Cir. 1991) (quoting *Whalen*, 429 U.S. at 599-600). As the Supreme Court has not yet ruled definitively on the issue, the federal judiciary has no unified view as to whether there is a constitutionally protected right to informational privacy. The majority of circuits considering this issue (the Second, Third, Fifth, and Ninth Circuits) find that there is. See, e.g., *Doe v. City of New York*, 15 F.3d 264, 267 (2d Cir. 1994) (concluding there is "a recognized constitutional right to privacy in personal information"); *Fadjo v. Coon*, 633 F.2d 1172, 1175-76 (5th Cir. 1981); *United States v. Westinghouse*, 638 F.2d 570, 577 (3d Cir. 1980), and *Roe v. Sherry*, 91 F.3d 1270, 1274 (9th Cir. 1996); *Doe v. Attorney General*, 941 F.2d at 795-96. A minority conclude there is not. See *J.P. v. DeSanti*, 653 F.2d 1080, 1090 (6th Cir. 1981).

[99]*United States v. Westinghouse Elec. Corp.*, 638 F.2d at 577 n.5.

individuals, such as dependents, retirees, and foreign nationals, to participate in officially sanctioned biometric programs. In such programs, individuals would be compelled to provide biometric identification information to the Army for collection, maintenance, use and dissemination in Army databases. Such Army-mandated use of biometrics implicates physical and informational privacy concerns and, to a lesser extent, decisional privacy concerns.[100]

Physical Privacy. *Constitutional challenges to fingerprinting in noncriminal context.* The overwhelming majority of the Army's biometric applications will fall into the noncriminal context, for such matters as network or physical security, fraud prevention, convenience, efficiency, etc. While the federal courts have not had occasion to rule on the government-mandated use of biometrics, many decisions have established that an individual has minimal constitutional privileges concerning his fingerprints.[101]

Moreover, the courts have upheld numerous federal, state, and municipal requirements mandating fingerprinting for employment and licensing purposes, provided that the government has a rational basis for requiring fingerprinting (*American Law Reports*, 1999, p. 732).[102] In a federal context, the so-called rational basis test means that Congress must show that the fingerprinting requirement bears a rational relationship to a legitimate government objective or interest.[103] For example, courts have upheld government-mandated

[100]The following hypothetical example might illustrate how decisional privacy concerns could be implicated by a biometric scheme. In response to growing concerns about missing children, the Army decides to require all children attending day care programs on Army bases to be biometrically scanned for identification purposes. Parents object on the grounds that they are fully satisfied with the less-intrusive security already offered at the day care programs on Army bases and that their children will be unduly traumatized by the scanning. Educational zone of privacy concerns are possibly implicated.

[101]See, e.g., *Davis v. Mississippi*, 394 U.S. 721, 727 (1969); *Schmerber v. California*, 384 U.S. 757, 764 (1966).

[102]See also, e.g., Department of Justice (1990, pp. 48–52).

[103]See, e.g., *Utility Workers Union of America, AFL-CIO, v. Nuclear Regulatory Commission*, 664 F.Supp. 136, 139 (S.D.N.Y. 1987); *Iacobucci v. City of Newport*, 785 F.2d 1354, 1355-56 (6th Cir. 1986), *rev'd on other grounds*, 479 U.S. 921 (1986); *Thom v. New York Stock Exchange*, 306 F.Supp. 1002, 1010 (S.D.N.Y. 1969). The rational basis test is a lesser standard of judicial scrutiny than the compelling state interest test. Courts apply the compelling state interest test when state action affects the exercise of

fingerprinting for employment and licensing purposes in connection with the taking of fingerprints for spouses of liquor licensees; male employees of alcoholic beverage wholesalers; taxi drivers; cabaret employees; bartenders; dealers in secondhand articles; all employees of member firms of national security exchanges registered with the Securities and Exchange Commission; and all individuals permitted unescorted access to nuclear power facilities.[104]

For example, in *Utility Workers Union of America v. Nuclear Regulatory Commission*, a union representing 5,170 utility workers in nuclear power plants challenged as unconstitutional that part of a newly enacted federal statute requiring that these workers be fingerprinted.[105] The union claimed the fingerprinting requirement violated the workers' Fourth Amendment and privacy rights. The federal district court disagreed and upheld the fingerprinting requirement. Citing a long string of cases, the court noted that in noncriminal contexts, the judiciary has "regularly upheld fingerprinting of employees."[106]

As for the constitutional right to privacy claim, the court quoted from a leading federal appellate court case:

> Whatever the outer limits of the right to privacy, clearly it cannot be extended to apply to a procedure the Supreme Court regards as only minimally intrusive. Enhanced protection has been held to apply only to such fundamental decisions as contraception . . . and family living arrangements. Fingerprints have not been held to merit the same level of constitutional concern.[107]

Moreover, in applying the rational basis test, the court noted congressional concern over an incident of sabotage at a nuclear power

---

a fundamental right, such as political speech. See, e.g., Department of Justice (1990, p. 48).

[104]See *American Law Reports* (1999, p. 732, citations omitted); *Utility Workers Union of America*, 664 F.Supp. at 136.

[105]*Utility Workers Union of America*, at 136. The union directed its challenge to Section 606 of the Omnibus Diplomatic Security and Anti-Terrorism Act of 1986, codified as section 149 of the Atomic Energy Act of 1954, 42 U.S.C. § 2169 (1986). 10 C.F.R. § 73.57 implements the statute.

[106]*Id.* at 138-39 (citations omitted).

[107]*Id.* at 139 (quoting *Iacobucci v. City of Newport*, 785 F.2d at 1357-58).

plant in Virginia and concluded that "[u]sing fingerprints to verify the identity and any existing criminal history of workers with access to vital areas or safeguards information is a rational method of clearing these workers."[108]

Similarly, in a case involving a challenge to a New York state regulation requiring fingerprinting of all employees of national stock exchanges, a federal district court found that "[p]ossession of an individual's fingerprints does not create an atmosphere of general surveillance or indicate that they will be used for inadmissible purposes. Fingerprints provide a simple means of identification no more." The court observed that as long as the government had a "valid justification . . . for the taking of the prints under reasonable circumstances, their use for future identification purposes even in criminal investigations, is not impermissible."[109]

Constitutional challenges to fingerprinting in a criminal justice context: What will happen when Army authorities want a biometric identifier from a member of the Army community whom they suspect has committed a crime? Capturing the biometric identifier in this context should not run afoul of the Constitution. The Fourth Amendment to the U.S. Constitution governs searches and seizures conducted by government agents. It provides that "[t]he right of the people to be secure in their persons, houses, papers, and effects, against unreasonable searches and seizures, shall not be violated." As the amendment makes clear, the Constitution does not forbid all searches and seizures, only "unreasonable" ones. The Supreme Court defines a search as an invasion of a person's reasonable expectations of privacy.[110] To evaluate whether providing a biometric identifier in a criminal justice context constitutes a search, the judiciary focuses on two factors. First, the court examines the nature of the intrusion.[111] Actual physical intrusions into the body, such as blood-drawing,[112] Breathalyzer testing, and urine analysis,[113] can

[108]*Utility Workers Union of America*, 664 F.Supp. at 139.

[109]*Thom v. New York Stock Exchange*, 306 F.Supp. at 1010.

[110]See, e.g., *Katz v. United States*, 389 U.S. 347, 360-62 (1967) (Harlan, J., concurring).

[111]See *Skinner v. Railway Labor Executives' Ass'n*, 489 U.S. 602, 616 (1989).

[112]See *Schmerber v. California*, 384 U.S. at 767-68.

[113]See *Skinner*, 489 U.S. at 618.

constitute Fourth Amendment searches. Second, the court examines the scope of the intrusiveness paying close attention to the "host of private medical facts" revealed during the search.[114]

In the criminal justice context, the Supreme Court has examined the issue of whether acquiring information about an individual's personal characteristics constitutes a search. It has found that requiring a person to give voice exemplars is not a search because "the physical characteristics of a person's voice, its tone and manner, as opposed to the content of a specific conversation, are constantly exposed to the public," such that "no person can have a reasonable expectation that others will not know the sound of his voice."[115] Using the same reasoning, the Court has ruled that requiring a person to give handwriting exemplars is not a search.[116] It has also described fingerprinting as nothing more than obtaining "physical characteristics . . . constantly exposed to the public,"[117] and that fingerprinting "involves none of the probing into an individual's private life and thoughts that marks an interrogation or search."[118]

In cases where provision of a biometric identifier might be found to constitute a search (as in the hypothetical case of a physically intrusive DNA-based biometric that would reveal extensive private medical facts about the individual), "the ultimate measure of the constitutionality of a governmental search is 'reasonableness.' "[119] To make this determination, a court must balance the "intrusion on the individual's Fourth Amendment interests against its promotion of legitimate governmental interests."[120] In the criminal context, a search is "reasonable" only if the law enforcement agency has probable cause of criminal activity.[121]

---

[114]See *id.* at 617.

[115]*United States v. Dionisio*, 410 U.S. 1, 93 (1973). See also LaFave et al., *Criminal Procedure*, Vol. 2, § 3.2(g) (LaFave et al., 1999).

[116]*United States v. Mara*, 410 U.S. 19, 93 (1973).

[117]*Cupp v. Murphy*, 412 U.S. 291 (1973).

[118]*Davis v. Mississippi*, 394 U.S. at 726-727. See also LaFave, *Criminal Procedure* at § 3.2(g).

[119]*Vernonia Sch. Dist. 47J v. Acton*, 515 U.S. 646, 652 (1995).

[120]*Id.* (quoting *Skinner*, 489 U.S. at 619) (internal quotation marks omitted).

[121]See *Zurcher v. Stanford Daily*, 436 U.S. 547, 554-55 (1978).

FBI experience: As the Army considers various biometric applications, it might benefit from study of the FBI experience involving the bureau's searchable criminal and civil fingerprint databases. In particular, the conclusion of the FBI's Office of the General Counsel (OGC) that the FBI's use for criminal justice purposes of fingerprint records obtained from servicemembers and federal employees is "legally unobjectionable" should be of interest to the Army.

Currently, individuals serving in the military service and those persons applying for federal employment must undergo fingerprinting.[122] While some of this fingerprinting is still done using the traditional ink-and-paper ten-print cards, much of it is being done electronically as a biometric, resulting in an image file, that can be transformed into a template. Eventually, all fingerprinting will be done through some type of biometric process.

By way of background, Congress in 1924 authorized the Department of Justice to begin collecting fingerprint and arrest record information voluntarily submitted for federal and state arrests. In 1930, Congress created the FBI's identification division, giving it responsibility for "acquiring, collecting, classifying, and preserving criminal identification and other crime records" and authorizing the exchange of these criminal identification records with authorized state, and local officials.[123] Today, the FBI's Criminal Justice Information Services (CJIS) Division is the world's largest fingerprint repository. Its current file holdings of fingerprint cards total over 219 million. This figure grows by over 5,000 each day (Archer, 1997).

The fingerprint records obtained from military members, federal applicants and others are submitted to the FBI. CJIS runs the finger-

---

[122]Executive Order 10450 (1953) requires federal employees in positions affecting the national security to submit fingerprints. Both DoD and the Army have Personnel Security Programs. These programs establish comprehensive policies and procedures applicable to personnel in the Army and other military branches; civilian employees in the DoD and Department of the Army; Army and DoD contractors; as well as other affiliated persons. Examples of affiliated persons include Red Cross or United Service Organizations personnel. The DoD Personnel Security Program and Army Personnel Security Program require the subject of each personnel security investigation to "provide fingerprints of a quality acceptable to the FBI" among other things. See 32 C.F.R. §§ 154.35, 154.8; AR 380-67, dated September 9, 1988.

[123]See *United States Department of Justice v. Reporters Committee for Freedom of Press*, 489 U.S. 749 (1989).

print record against its integrated, automated "criminal files" database in Clarksburg, West Virginia, to determine if the individual has any past criminal involvement.[124] CJIS receives about 32,000 fingerprint cards a day for such processing.[125] This database contains a comprehensive fingerprint record of individuals fingerprinted after arrest or incarceration (FBI, 1995).[126] CJIS has converted all of these fingerprint records into electronic format. Moreover, state and local criminal justice agencies have the capability to transmit their fingerprint records to CJIS electronically. Thus, the "criminal files" database is readily and easily searchable (FBI, 1999).[127]

If the criminal history background search reveals an individual's past criminal involvement, the fingerprint record becomes part of the "criminal files" database. This database has more than 132 million criminal cards representing 36.1 million individuals who have been arrested or convicted of a criminal offense in the United States.

If the search reveals no criminal history, the fingerprint record is kept in the CJIS "civil files" database (FBI, 1995). The "civil files" database maintains approximately 87 million civil fingerprint cards represent-

---

[124]See, e.g., FBI (1995). This process is sometimes referred to as a National Agency Check (NAC). See e.g., AR 380-67.

[125]This number includes criminal history background searches requested by federal and state governments and others for various permit, license and employment clearances in addition to federal employment applications and military service. See FBI (1995).

[126]See also *Identification Division Records System Notice*, printed in 55 Fed. Reg. 49174 (Vol. 55, No. 227, November 26, 1990).

[127]On August 10, 1999, FBI Director Louis J. Freeh inaugurated the full operation of the Integrated Automated Fingerprint Identification System (IAFIS), which provides federal, state, and local criminal justice agencies the ability to transmit fingerprint information electronically. Previously, criminal justice agencies mailed ink and paper fingerprint cards to the FBI for processing. After the cards were received, a semiautomated system classified the fingerprints and compared them to the fingerprint cards in the FBI's CJIS fingerprint database. This identification process sometimes took weeks to complete. IAFIS will compare the submitted images with its huge database of fingerprints, and respond within two hours. The response will include a complete criminal history of the person, if one exists. Even if the person fingerprinted provides false identification, IAFIS will make a positive identification by matching fingerprints. For a discussion of law enforcement use of such automated systems, see Garfinkel (2000, pp. 41–46) (in his book, Garfinkel expresses concerns about biometric technologies eroding privacy).

ing approximately 39 million people.[128] These individuals have been fingerprinted as a result of federal employment applications or military service, for alien registration and naturalization purposes, as well as for voluntary submission for personal identification purposes.[129]

The "civil files" database is not fully automated. The vast bulk of its fingerprint records are the paper and ink variety. However, since last year, the FBI has taken steps to automate the database from "Day One forward" as it receives biometric versions, i.e., new fingerprint cards in electronic (biometric) format (CJIS, 1998). The FBI also has the option of scanning into the database the paper and ink records to convert them into electronic format. The FBI could also electronically organize subfiles, known as Special Latent Cognizant (SLC) files, within the "civil files" database. For example, the FBI could organize an SLC file of fingerprints of DoD employees and military members. By getting civil fingerprint records electronically recorded into the "civil files" database and by organizing extensive SLC files, or subsets, within it, the "civil files" database, like the "criminal files" database, would be easily and readily searchable.

From a criminal investigative point of view, the capability to access and search latent fingerprints against all the fingerprint records in the "civil file" database would be of great benefit to law enforcement. For example, a latent fingerprint found at a crime scene on a military base could be searched against the DoD SLC file in the "civil files" database. In this way, more crimes could be solved.

In 1995, the FBI asked its OGC for its legal opinion as to whether the FBI could conduct such searches of its "civil files" database. After review, OGC concluded that "[u]sing civil fingerprint records for criminal justice purposes is legally unobjectionable" (FBI, 1995).[130] OGC determined that the use of CJIS "civil files" for criminal justice purposes is consistent with the Privacy Act because it is a routine

---

[128]See, e.g., CJIS (1998). Some people have more than one fingerprint record in the database. For example, many military veterans take employment with the federal government.

[129]See, e.g., CJIS (1998). The number of fingerprint records from voluntary submissions is very small.

[130]Please note: This OGC opinion does not have the force of law.

use. The two requirements for routine use are compatibility with the original use for which the data were collected and *Federal Register* publication.[131] OGC determined that the compatibility requirement is met because "using fingerprints collected for criminal history check purposes for criminal justice identification purposes is . . . completely compatible with the purposes for which they were first collected" (FBI, 1995). OGC also determined that because the FBI has properly published the routine use in the *Federal Register,*[132] "after reading the notice, no reasonable person could claim to be surprised to find that [his] fingerprints, once submitted to the FBI, will be used by the Bureau for identification purposes in either a criminal justice or civil setting" (FBI, 1995). Similarly, drawing on some of the case law discussed above, OGC determined that no constitutional right to privacy exists in an individual's identity and criminal history background.[133]

In sum, OGC has concluded that it is legally unobjectionable for the FBI to search its "civil files" database, which it can organize into SLC

---

[131]5 U.S.C. § 552a(e)(4)(D).

[132]The FBI changed the routine uses set forth in the "Fingerprint Identification Record System" notice in February 1996. The notice reads, in pertinent part:

> Identification and criminal history record information within this system of records may be disclosed as follows:
>
> > To a Federal, State, or local law enforcement agency, or agency/organization directly engaged in criminal justice activity (including the police, prosecution, penal, probation/parole, and the judiciary), and/or to a foreign or international agency/organization, consistent with international treaties, conventions, and/or executive agreements, *where such disclosure may assist the recipient in the performance of a law enforcement function, and/or for the purpose of eliciting information that may assist the FBI in performing a law enforcement function; to a Federal, State, or local agency/organization for a compatible civil law enforcement function; or where such disclosure may promote, assist, or otherwise serve the mutual criminal law enforcement efforts of the law enforcement community. . . .*

See 61 Fed. Reg. 6386 (Vol. 61, No. 34, February 20, 1996) (emphasis added).

[133]See FBI (1995) (quoting *Trade Waste Mgt Ass'n v. Hughey,* 780 F.2d 221, 234 (3d Cir. 1985), "While it may be that when conduct resulting in the convictions or charges was engaged in the person who engaged in it expected that such participation would remain secret, that expectation was never reinforced by law.")

subsets. This conclusion suggests that it would be legally unobjectionable for the Army, were it so inclined on policy grounds, to coordinate with the FBI to have the FBI organize an SLC file consisting of the overall Army community, or various SLC files containing further subsets of the Army community, such as active-duty and Department of the Army civilians, which the FBI could then search. The OGC conclusion further suggests that it would be legally unobjectionable for the Army, were it so inclined on policy grounds, to organize its own similar "civil" database of biometric identification information and search this database for criminal justice identification purposes, provided that the Army received the proper authority to do so and complied with the Privacy Act requirements, as the FBI did. No apparent constitutional barriers stand in the way. Before embarking on these database paths, however, the Army should undertake a detailed legal analysis based on exactly what it wants to do to make certain it is on firm legal ground. In addition, the Army will have to assess policy concerns related to such uses.

DMDC experience: The Defense Manpower Data Center (DMDC) operates what is arguably DoD's largest biometric database. DMDC's experiences in this regard might be instructive to the Army. (The DMDC experience is also included in Appendix B.)

The Federal Managers' Financial Integrity Act of 1982 requires federal managers to establish internal controls to provide reasonable assurance that funds, property and other assets are protected against fraud or other unlawful use. As a result of this legislation, DoD launched Operation Mongoose, a fraud prevention and detection initiative. Operation Mongoose exposed a number of fraud schemes and indicated that DoD needed to improve servicemember identification and verification procedures. Responding to the need for better fraud prevention measures, the Acting Under Secretary of Defense (Personnel and Readiness) gave authority to the DMDC to initiate an electronic fingerprint capture policy in 1997.[134]

Since 1998, the DMDC has been capturing the right index fingerprint of all active-duty, reserve, and retired military personnel as well as survivors receiving a military annuity. This potential enrollment

---

[134]See Finch (1997) and, generally, Harreld (1999).

pool is some three million people. The print is captured during routine issuance (or reissuance) of military ID cards at some 900 DMDC sites. DMDC stores electronic copies of these fingerprints (the file images and the templates) in a comprehensive database, the Defense Enrollment Eligibility Reporting System (DEERS). DMDC stores no copies of fingerprints on the actual military ID card. DEERS can be accessed if a person's identity needs to be authenticated.

The DEERS database is believed to be the largest biometric database in DoD. As such, it complies with the Privacy Act. The Defense Privacy Office and other institutional assets assisted in ensuring compliance. To date, no successful legal challenge has been brought against the DMDC's biometric database.

Information Privacy—*Whalen v. Roe.*[135] *Introduction to Whalen v. Roe*: The Supreme Court's 1977 decision in *Whalen v. Roe* "began the process of identifying the elements of an American constitutional right of informational privacy" (Schwartz, 1995).[136] In 1999, a federal court cited *Whalen* for the proposition that "the Constitution protects an individual's privacy right to avoid disclosure of personal information."[137]

Whalen involved the constitutional question of whether the state of New York could record and store, in a centralized computer database, "the names and addresses of all persons who have obtained, pursuant to a doctor's prescription, certain drugs."[138]

---

[135]This section of the report is largely drawn from Woodward (1997a).

[136]Other legal scholars have perhaps interpreted the significance of *Whalen v. Roe* slightly differently. See, e.g., Allen (1991, p. 181), "The Court has come closest to recognizing an independent right of information privacy in *Whalen v. Roe*"; Roch (1986, pp. 71, 89), "[I]n *Whalen v. Roe*, the court [sic] recognized in dicta that there may exist a right to protect against improper disclosure of personal data."; Cate (1997, p. 63), "[H]aving found this new privacy interest in nondisclosure of personal information, the Court . . . applying a lower level of scrutiny, found that the statute did not infringe the individual's interest in nondisclosure"); and Strauss et al. (1995, p. 874), "A requirement that information of arguable utility to a lawful regulatory program be collected or submitted is unlikely to fall beyond the constitutional power of either federal or state government."

[137]*Wilson v. Pennsylvania State Police*, CA 94-6547, 1999 U.S. Dist. LEXIS 3165 *5 (E.D. Pa. March 11, 1999) (U.S. Mag. Judge Hart) (citing *Whalen v. Roe*, 429 U.S. at 599-600). See also *In re Crawford*, 1999 U.S. App. LEXIS 24941 at *16.

[138]*Whalen v. Roe*, 429 U.S. at 591.

While technology has changed greatly since 1977, the legal reasoning in *Whalen* is still relevant, particularly for biometrics, and more important, for the Army's use of biometrics. *Whalen* is instructive because it demonstrates the federal judiciary's likely approach to deciding some of the major constitutional law issues likely to be raised by Army-mandated biometric applications. Accordingly, the facts of the case, the holding, and the judicial reasoning deserve detailed examination.

Facts: In 1970, the New York legislature, disturbed about the growing drug problem, established a commission to evaluate the state's drug control laws.[139] After study, the commission made recommendations to correct perceived deficiencies in these state laws. Following up on these recommendations, the legislature amended the New York Public Health Law to require that "all prescriptions for Schedule II drugs" had to be prepared by the physician on an official state-provided form.[140] The completed form identified

- the prescribing physician;
- the dispensing pharmacy;
- the prescribed drug and prescribed dosage; and,
- the name, address, and age of the patient.

The statute required that a copy of the completed form be forwarded to the New York State Department of Health in Albany.[141] Albany received about 100,000 Schedule II prescription forms each month. There, the government agency recorded the information on magnetic tapes for eventual processing by computer. Based on his study of other states' reporting systems, the commission's chairman found

---

[139]See *id.* This commission was formally known as The Temporary State Commission to Evaluate the Drug Laws. See *id.* at 592 n.4.

[140]*Id.* at 593. The statute classified potentially harmful drugs in five schedules which conformed with relevant federal law. Schedule II drugs included the most dangerous of the legitimate drugs. Examples of such drugs would include opium, methadone, amphetamines, and methaqualone. These drugs all have accepted medical uses. The statute also provided for an emergency exception.

[141]The office which received the forms was the Bureau of Controlled Substances, Licensing and Evaluation. See *Roe v. Ingraham*, 403 F. Supp. 931, 932 (S.D.N.Y. 1975), *rev'd, Whalen v. Roe*, 429 U.S. 589 (1977).

that this comprehensive government-mandated database would serve two purposes: It would be a "useful adjunct to the proper identification of culpable professionals and unscrupulous drug abusers," and it would provide the authorities a "reliable statistical indication of the pattern of [the state's] drug flow" to help stop the diversion of lawfully manufactured drugs into the illegal market.[142]

Patients, doctors, and physicians' associations challenged the New York statute in court. The evidence offered before the federal district court, where the case was first heard, included testimony from

- two parents who "were concerned that their children would be stigmatized [as drug addicts] by the State's central filing system";
- three adult patients who "feared disclosure of their names" to unauthorized third parties; and,
- four physicians who believed that the New York statute "entrenches on patients' privacy and that each had observed a reaction of shock, fear, and concern on the part of their patients."[143]

The parties thus advanced two related privacy concerns, which eventually reached the Supreme Court's consideration: "the nondisclosure of private information," or informational privacy, and an individual's "interest in making important decisions independently," or decisional privacy.

Holding: In his opinion for the Court, Justice John Paul Stevens, joined by the Chief Justice and five other justices, found that "neither the immediate nor the threatened impact of the [statute's] patient-identification requirements . . . on either the reputation or the independence of patients . . . is sufficient to constitute an invasion of any right or liberty protected by the Fourteenth Amendment."[144] With these words, the Supreme Court rejected the privacy claim. In sum, the nation's highest court ruled that a government's centralized,

---

[142] *Whalen v. Roe,* 429 U.S. at 592 n.6.

[143] *Id.*

[144] *Id.* at 603-04.

computerized database containing massive amounts of sensitive medical information passed constitutional muster.

Judicial reasoning: What factors influenced the Supreme Court's reasoning? First, the Court seemed impressed by the New York legislature's creation of a specially appointed commission that held many hearings on and conducted a thorough study of the state's drug problem.[145] The commission consulted extensively with authorities in other states that used central reporting systems effectively. In other words, a commission empowered by the legislature had done its homework in an attempt to help solve the menacing problem of drugs. The Court concluded that the statute was "manifestly the product of an orderly and rational legislative decision."[146]

Arguably, the New York statute had not had much of an impact on the drug problem. For example, 20 months after its enactment, examination of the database led to only two investigations involving illegal use of drugs. As a kind of political process check, the Court explained that the state legislature, which gave this patient identification requirement its legal life, can also sound its death knell if it turns out to be an "unwise experiment."

In its analysis of the informational privacy concerns raised, the Court paid close attention to what specific steps the state agency had taken to prevent any unauthorized disclosures of information from the centralized database. In particular, the Court noted the following:

- The forms and records were kept in a physically secure facility.
- The computer system was secured by restricting the number of computer terminals that could access the database.
- Employee access to the database was strictly limited.
- There were criminal sanctions for unauthorized disclosure.

The Supreme Court took a somewhat practical approach to the way personal information is used in the contemporary age. It accepted

---

[145]See *id.* at 591. The Temporary State Commission to Evaluate the Drug Laws issued two reports, which the legislature made part of the legislative history of the statute.

[146]*Id.* at 597.

the view that disclosure of such medical information to various government agencies and private sector organizations, such as insurance companies, is "often an essential part of modern medical practice even when the disclosure may reflect unfavorably on the character of the patient. Requiring such disclosures to representatives of the State having responsibility for the health of the community does not automatically amount to an impermissible invasion of privacy."[147]

In addressing decisional privacy issues, the Court acknowledged genuine concern that the very existence of the database will disturb some people so greatly that they will refuse to go to the doctor to get necessary medication. However, given the large number of prescriptions processed at Albany, the Court came to the conclusion that the "statute did not deprive the public of access to the [legal] drugs."[148]

The Court's opinion concluded with a cautionary note that still echoes loudly 23 years later:

> We are not unaware of the threat to privacy implicit in the accumulation of vast amounts of personal information in computerized data banks or other massive government files. . . . The right to collect and use such data for public purposes is typically accompanied by a concomitant statutory or regulatory duty to avoid unwarranted disclosures.[149]

The New York statute and its related implementation showed "a proper concern with, and protection of, the individual's interest in privacy."[150] The Court, however, limited the effect of its decision by reserving for another day consideration of legal questions that could arise from unauthorized disclosures of information from a

[147]*Id.* at 602 (footnote omitted).

[148]*Id.* (noting that Albany received approximately 100,000 prescription forms for Schedule II drugs monthly).

[149]*Id.* at 605.

[150]*Id.*

government database "by a system that did not contain comparable security provisions."[151]

Justice Brennan's concurring opinion: In his concurring opinion, Justice William Brennan, more so than his colleagues, expressed his concern over the potential erosion of informational privacy in the face of emerging technologies. "The central storage and easy accessibility of computerized data vastly increase the potential for abuse of that information, and I am not prepared to say that future developments will not demonstrate the necessity of some curb on such technology."[152] While this specific "carefully designed program" did not "amount to a deprivation of constitutionally protected privacy interests," Justice Brennan suggested that there is a core right to informational privacy and stressed that future programs might be subjected to a compelling state interest test or strict scrutiny by the court of the government action.[153]

Justice Stewart's concurring opinion: Justice Potter Stewart, in his concurrence, took issue with what he implicitly viewed as Justice Brennan's expansive privacy approach as well as with his brethren's view of constitutional privacy interests. According to Stewart, no general right of privacy can be found in the Constitution. Moreover, in Stewart's view, privacy concerns are matters left largely to the individual states.[154]

Cautionary note: The *Whalen* Court expressed its concern about the potential for "unwarranted disclosures" from the government's databases. As Professor Bevier, writing in a similar context, has explained:

> The fact that the government collects such great quantities of data gives rise to concern . . . that the data will be inappropriately disseminated, within government or to outsiders, or that it will be otherwise misused or abused. Recent advances in computer technology, which permit data to be manipulated, organized, compiled,

---

[151]*Id.* at 605-06.

[152]*Id.* (Brennan, J., concurring).

[153]See *id.* (Brennan, J., concurring).

[154]See *id.* (Stewart, J., concurring).

> transferred, distributed, and retrieved with hitherto unimaginable ease, exacerbate such concern. (Bevier, 1995, p. 457.)[155]

With the exception of Justice Stewart, all of the justices adopted a prospective approach. That is, by intensely focusing on the facts of *Whalen*, the Court left itself ample judicial wiggle room to find that government-mandated use of new technologies combined with powerful computer systems might lack necessary constitutional safeguards. Because the *Whalen* decision is tied so intimately to the specifics of *Whalen*, a future Court could easily distinguish the facts of a future case from the facts of *Whalen* to reach a different decision.

In sum, a lesson for the Army to take away from *Whalen* is that a future Court might find an informational privacy right violated unless the government agency collecting the information (1) had made clear its need and purpose in collecting the information and (2) had taken strong and effective measures to prevent unwarranted disclosures from its databases. In other words, if the government agency ignores these steps, the Court's cautionary note of *Whalen* could turn into a clear-sounding constitutional alarm bell in the future.[156]

The legacy of Whalen: Recent case law suggests that the federal judiciary accepts the informational privacy concept articulated in *Whalen*. In 1999, the Court of Appeals for the Ninth Circuit explained that one of the constitutionally protected privacy interests of *Whalen* is "the individual interest in avoiding disclosure of personal matters."[157] Moreover, the Ninth Circuit, like the *Whalen* court, found that the right to informational privacy is not absolute but must be balanced with the governmental interest.

In *In re Crawford*, the court held that federally required public disclosure of the SSNs of certain paralegals does not violate any consti-

---

[155]See also Garfinkel (2000, pp. 260–266).

[156]For this observation, the principal author thanks Professor Steve Goldberg of the Georgetown University Law Center who shared it in September 1996. Professor Goldberg explained that when a Supreme Court opinion offers broad pronouncements and little factual analysis, it is a sure sign that the Court is on comfortable turf. However, when the opinion deals with intense factual scrutiny, the Court is less sure of itself and thus keeping its options open for the long run.

[157]*In re Crawford*, 1999 U.S. App. LEXIS 24941 at *7-8.

tutional or statutory rights of these individuals. The federal law at issue requires a bankruptcy petition preparer (BPP), a type of paralegal, to include his or her SSN on all documents filed with the federal bankruptcy courts. By law, these documents are public records that can be accessed by anyone. Jack Ferm, a BPP, refused to provide his SSN on bankruptcy documents he had filed with a bankruptcy court in Nevada. He feared disclosure of his SSN would make him particularly vulnerable to the crime of identity theft. When the court fined him for refusing to provide his SSN, Ferm filed a lawsuit in federal court, claiming the disclosure of his SSN violated his privacy rights.

Although the court sympathized with Ferm's "speculative fear," it noted that an SSN, "unlike HIV status, sexual orientation, or genetic makeup" is "not inherently sensitive or intimate information, and its disclosure does not lead directly to injury, embarrassment, or stigma."[158] The court balanced Ferm's interest in nondisclosure of his SSN with the governmental interests. The many factors the court considered included:

> [T]he type of record requested, the information it does or might contain, the potential for harm in any subsequent nonconsensual disclosure, the injury from disclosure to the relationship in which the record was generated, the adequacy of safeguards to prevent unauthorized disclosure, the degree of need for access, and whether there is an express statutory mandate, articulated public policy, or other recognizable public interest militating toward access.[159]

The court found that the disclosure requirement serves the Bankruptcy Code's "public access" provision, which is rooted in the traditional right of public access to judicial proceedings. After weighing the many relevant factors, the court concluded:

> [T]he speculative possibility of identity theft is not enough to trump the importance of the governmental interests [requiring public disclosure]. In short, the balance tips in the government's favor.

---

[158] *Id.* at *13-14. In Ferm's case, he had not suffered any actual identity theft at the time he brought his suit, thus the court determined his fear as "speculative."

[159] *Id.* at *11 (citing *Doe v. Attorney General*, 941 F.2d at 796, quoting *Westinghouse*, 638 F.2d at 578).

> Accordingly, we cannot say that Congress transgressed the bounds of the Constitution in enacting the statutes at issue here.[160]

In re Crawford is a recent example illustrating that the Supreme Court's approach in *Whalen* remains firmly in place within the federal judiciary. It is prudent for the Army to study *Whalen* closely, to explain its military need for biometrics and to have database safeguards in place.

What Constitutionally Based Religious Concerns Does Biometrics Raise?

Overview. As explained above, some limited segments of American society have expressed religious objections to the use of biometrics. Among these objections, individuals oppose being compelled to participate in a government-mandated biometric application. The New York Department of Social Services and the Connecticut Department of Social Services (DSS) have encountered legal challenges based on religious concerns from entitlement program recipients who refused to provide a biometric identifier. Based on these objections, the Army might encounter a similar legal challenge to its mandated use of biometrics. Accordingly, the New York DSS and Connecticut DSS experiences might offer useful insights for the Army.

New York Experience. Liberty Buchanan, a New York resident, received AFDC and Food Stamps for herself and her four minor children. In 1996, New York DSS told her she would be required to participate in an AFIS. New York law required participation in AFIS as a condition of eligibility for AFDC and other entitlements.[161] Buchanan refused to participate in AFIS. She based her refusal on her religious convictions, grounded in part on her interpretation of the "mark of the beast" language in the Book of Revelation. Because she refused to provide a fingerprint, DSS discontinued the family's benefits. After a DSS agency hearing, the State Commissioner of Social Services affirmed the DSS decision, finding that Buchanan did not demonstrate a good cause basis for exemption from the finger

---

[160] *Id.* at *16. The court did, however, "encourage the Bankruptcy Courts to consider enacting rules to limit the disclosure of BPP SSNs."

[161] See 18 N.Y.C.R.R. § 351.2(a).

imaging requirement. Buchanan then appealed to the New York state court. After a hearing, the court found that Buchanan had failed to "set forth any competent proof that the AFIS actually involved any invasive procedures marking them in violation of [her] beliefs."[162] Accordingly, the court upheld the DSS decision.

Connecticut Experience. Similarly, in Connecticut, John Doe, his wife, and minor children—recipients of Temporary Family Assistance (TFA)—refused to submit to the Connecticut DSS digital imaging requirement.[163] Beginning in January 1996, DSS, pursuant to state law, began requiring all TFA recipients to be biometrically imaged for identification purposes by providing copies of the fingerprints of their two index fingers (*Uniform Policy Manual*, 2000).[164] In April 1996, Mr. and Mrs. Doe objected, based on their religious beliefs. DSS exempted them from the requirement in April 1996 and October 1997. In July 1998, however, DSS reviewed its policy and determined that the Does would have to comply with the digital imaging requirement. Doe requested a DSS hearing.

At the August 1998 hearing, he testified about his objections to providing a biometric identifier. He based these objections on his religious beliefs. Doe testified that the Book of Revelation discusses the "mark or number of the beast," which the "beast" tries to make all people receive on their hand or forehead. According to Doe, those who accept the mark "shall drink of the wine of the wrath of God" and be condemned. By submitting to digital imaging and allowing himself to be marked in this way, he would violate his religious convictions. He therefore requested a "good cause" exception to the digital imaging requirement as provided in the DSS regulations.[165]

In November 1998, the hearing officer ruled that Doe, "although having strong religious beliefs, some of which he interprets as a bar-

---

[162] *Buchanan v. Wing*, New York Supreme Court, Appellate Division, Third Judicial Department, December 4, 1997, 79341.

[163] "John Doe" is an alias used to protect the true identity of the individual out of respect for his and his family's privacy.

[164] See also Connecticut State DSS (1996, 2000).

[165] *Uniform Policy Manual* (2000), "Good cause is considered to exist when circumstances beyond the individual's control reasonably prevent participation with the Digital Imaging process."

rier for him to be digitally imaged, does not have as a result of this religious belief a circumstance beyond his control which prevents him from being digitally imaged" (Connecticut State DSS, 1998). Doe appealed from this final DSS decision to the Connecticut state court. While his case was pending, the DSS Commissioner decided to vacate the hearing decision and grant the Does an exception from the digital imaging requirement (Connecticut State DSS, 1999).

***Goldman v. Weinberger*.** While the Army has not yet encountered any legal challenges to its biometric applications, the U.S. military has encountered objections to military regulations based on an individual's religious beliefs. One of the best-known legal challenges brought against the military on this basis is the case of *Goldman v. Weinberger*, decided by the Supreme Court in 1986.[166]

S. Simcha Goldman, an Air Force officer and ordained rabbi of the Orthodox Jewish faith, was ordered not to wear his yarmulke while on duty and in uniform, pursuant to Air Force regulations.[167] Goldman then brought an action in federal district court, claiming that the application of the Air Force regulation to prevent him from wearing his yarmulke infringed upon his First Amendment freedom to exercise his religious beliefs. The District Court agreed with Goldman and permanently enjoined the Air Force from enforcing the regulation against him. The Court of Appeals reversed, and Goldman appealed to the Supreme Court.

The Supreme Court held that Goldman's religious objections, grounded in the First Amendment's free exercise of religion clause, did not prohibit the challenged regulation from being applied to Goldman, even though its effect is to restrict the wearing of the headgear required by his religious beliefs. The Court found that the First Amendment does not require the military to accommodate such practices as wearing a yarmulke in the face of the military's view that such practices would detract from the uniformity sought by dress regulations. In his majority opinion, then-Justice Rehnquist

---

[166]*Goldman v. Weinberger*, 475 U.S. at 503.

[167]Air Force Regulation 35-10 provides, in pertinent part, that authorized headgear may be worn out of doors but that indoors "[h]eadgear [may] not be worn . . . except by armed security police in the performance of their duties." AFR 35-10, ¶ 1-6.h(2)(f) (1980).

explained that, "when evaluating whether military needs justify a particular restriction on religiously motivated conduct, courts must give great deference to the professional judgment of military authorities concerning the relative importance of a particular military interest."[168]

Congress reacted to the *Goldman* decision by passing a statute effectively eviscerating the Court's ruling. In 1987, Congress amended the U.S. Code to permit a member of the armed forces to "wear an item of religious apparel while wearing the uniform of the member's armed force," with two exceptions: when "wearing of the item would interfere with the performance of the member's military duties" or if "the item of apparel is not neat and conservative."[169]

In 1990, the Supreme Court decided another important case involving religious beliefs. In *Employment Division, Department of Human Resources of Oregon v. Smith,* ("*Smith*"), Alfred Smith and Galen Black brought suit against the Oregon State Employment Division after it refused their claims for unemployment compensation.[170] Their employer had discharged them from their jobs on "misconduct" grounds because they had ingested peyote, a hallucinogen, as part of the sacramental observances of their Native American religion. Under Oregon law, peyote is a controlled substance and thus prohibited.

In *Smith,* the Supreme Court held that the First Amendment's free exercise of religion clause does not require exemption from a religiously neutral law for those whose religious beliefs preclude them from complying with the law. *Smith* holds that the legislature is free, however, to grant religious exemptions to the neutral laws if it so chooses.[171] Thus, in *Smith,* the First Amendment's free exercise clause did not prohibit the application of Oregon state drug laws to use of peyote for religious purposes. However, were it so inclined, the Oregon state legislature could create a religious exemption.

---

[168] *Goldman v. Weinberger,* 475 U.S. at 507 (citation omitted).

[169] See 10 U.S.C. § 774 (1999).

[170] *Employment Division, Department of Human Resources of Oregon v. Smith,* 494 U.S. 872 (1990).

[171] *Id.* at 889.

Lessons Learned. As it plans its biometric applications, the Army can draw several broad lessons from *Goldman*. First, *Goldman* demonstrates that just as the Supreme Court deferred to the Air Force uniform regulation, the federal judiciary will be somewhat deferential to an Army biometric application. Second, the congressional reaction to the *Goldman* decision demonstrates that Congress is not unwilling to require the military to make special allowances for religious objections of members of the military community. Third, the military, as an institution, and the Army, as an armed service, know that they take orders from Congress.

In the context of the Army's use of biometrics, the lesson from *Smith* reinforces a lesson from *Goldman*: While the Army's requirement for participating in biometric applications, just like the Oregon law prohibiting peyote, will be religiously neutral, Congress, like the Oregon state legislature, could grant, if it were so inclined, religious exemptions to the neutral requirement.

The Army, for example, has an extensive, established policy in place to accommodate religious practices.[172] It approves requests for accommodation of religious practices unless the accommodation will have an adverse impact on "military necessity," which consists of unit and individual readiness, unit cohesion, morale, discipline, safety, or health. As the Army's primary advisor on matters pertaining to religious accommodation, the Army Chief of Chaplains is an important institutional asset on whom the Army leadership may call for guidance in determining how religious objections to biometric applications should be handled. As the official charged with establishing the Army's policy on the accommodation of religious practices, the Deputy Chief of Staff for Personnel (DCSPERS) will also play a key role.

INTERNATIONAL LAW CONCERNS

European Union Data Protection Directive

Overview. The European Union Privacy Directive, also known as the EU Data Protection Directive or Directive 96/46/EC, took effect on

---

[172]See AR 600-20, ¶ 5-6 "Accommodating religious practices," dated July 15, 1999.

October 25, 1998.[173] The directive prohibits the transfer of personal data to any country that does not provide an "adequate" level of protection, as determined by the EU, for the privacy of the data. To ensure compliance with this adequacy requirement, all EU member states were obligated to enact comprehensive privacy legislation, by the effective date of the directive, requiring organizations to implement personal data policies. The United States, however, does not rely on this type of comprehensive legislation to protect privacy, but instead uses a "sectoral approach," relying on a combination of legislation, regulation, and self-regulation. These differing approaches to protecting privacy created uncertainty as to the impact on U.S. organizations of the directive's adequacy requirement.[174]

To address these concerns, the United States and the European Commission agreed in July 2001 on a "safe harbor" framework, under which eligible U.S. organizations can satisfy the "adequacy" requirements of the directive by voluntarily adhering to a set of data protection principles.[175]

Major Provisions. The directive has the potential to be far-reaching. For example, the EU personal data policies provide for the following:

- Transparency: Data must be processed fairly and lawfully.
- Purpose Limitation: Data must be collected and possessed for specified, legitimate purposes and kept no longer than necessary to fulfill the stated purpose.
- Data Quality: Data must be accurate and up-to-date.

---

[173]The official name of the directive is Directive 95/46/EC of the European Parliament and of the Council of 24 October 1995 on the protection of individuals with regard to the processing of personal data and on the free movement of such data, available at http://www2.echo.lu/legal/en/dataprot/directiv/directiv.html. For a comprehensive analysis of the EU Privacy Directive, see Swire and Litan (1998).

[174]See Safe Harbor Privacy Principles Issued by the U.S. Department of Commerce, dated July 21, 2001, available at http://www.export.gov/safeharbor/SHPRINCIPLES FINAL.htm.

[175]See Commission Decision Pursuant to Directive 05/46/EC of the European Parliament and of the Council on the adequacy of the protection provided by the Safe Harbor Privacy Principles and related Frequently Asked Questions, issued by the U.S. Department of Commerce, available at http://europa.eu.int/comm/internal_market/en/media/dataprot/news/decision.pdf.

- Data Transfers: Article 25 of the directive restricts authorized users of personal information from transferring that information to third parties without the permission of the individual providing the data, or data subject. In the case of data transfers across national boundaries, the directive prohibits data transfers outright to any country lacking an "adequate level of protection," as determined by the EU. Article 25 is a major source of U.S. concern.
- Special Protection for Sensitive Data: This provision requires restrictions on, and special government scrutiny of, data collection and processing activities of information identifying "racial or ethnic origin, political opinions, religious or philosophical beliefs, . . . [or] concerning health or sex life." Under the directive, such data collection or processing is generally forbidden outright.
- Government Authority: Each EU member state must create an independent public authority to supervise personal data protection. The EU will oversee the directive's implementation and will engage in EU-level review of its provisions.
- Data Controllers: Organizations processing data must appoint a "data controller" responsible for all data processing, who must register with government authorities.
- Individual Redress: A data subject must have the right to access information about himself, correct or block inaccuracies, and object to information's use.

Article 1 of the directive requires member states to protect the "fundamental rights and freedoms of natural persons, and in particular their right to privacy with respect to the processing of personal data." In essence, the EU has made privacy a fundamental human right.

Applicability to Biometrics. The directive defines personal data as any information relating to an identified or identifiable natural person. An identifiable person is one who can be identified, directly or indirectly, in particular by reference to an identification number or to one or more factors specific to his physical, physiological, mental, economic, cultural, or social identity. While the word "biometric" is not specifically cited in the text, biometric identifiers will likely be implicated by the directive's definition of personal data.

Applicability to the U.S. Army. If a U.S. organization wishes to receive personal data from an EU organization—for example, if the U.S. Army wishes to collect biometric finger images from its employees, including foreign nationals, at a base in Germany—the U.S. organization can comply with the directive in three ways. It can either join the safe harbor, satisfy one of the directive's other exceptions, or seek an adequacy determination (U.S. Department of Commerce, 2001).

If an organization decides to participate in the safe harbor, it must comply with the safe harbor requirements, which are set forth in a set of seven privacy principles, and it must publicly declare its adherence to these principles. This "self-certification" of compliance must be made annually to the U.S. Department of Commerce, which will maintain a regularly updated list of safe harbor participants through its Web site.[176] With regard to enforcement, organizations must have in place compliance verification procedures, as well as dispute-resolution mechanisms to resolve complaints about compliance. Further enforcement is provided for under U.S. federal or state law governing unfair and deceptive acts (U.S. Department of Commerce, 2001).

Accordingly, to be eligible to join the safe harbor, a U.S. organization must be subject to the jurisdiction of specified government bodies in the United States.[177] The Federal Trade Commission and the Department of Transportation are the only two such government bodies specified in the safe harbor agreement. As such, the U.S. Army would seem to be ineligible to join the safe harbor at this time, although the safe harbor agreement makes provision for its "review

---

[176]For the seven privacy principles, as well as the guidance for the implementation of these principles contained in the Frequently Asked Questions, see U.S. Department of Commerce (2001).

[177]See Article 1(2)(b) of Commission Decision Pursuant to Directive 05/46/EC of the European Parliament and of the Council on the adequacy of the protection provided by the Safe Harbor Privacy Principles and related Frequently Asked Questions, issued by the U.S. Department of Commerce, available at http://europa.eu.int/comm/internal_market/en/media/dataprot/news/decision.pdf.

in light of experience," such that the number of those organizations eligible to join the safe harbor may expand in the future.[178]

Even if the Army cannot avail itself of the safe harbor route to compliance with the directive, it appears very likely that the Army's use of biometrics will fit within one of the several exceptions and exemptions contained in the directive. Prominent among them, Article 3(2) of the directive contains an exception for public security and defense matters. It is not clear, however, whether this exception would be interpreted to apply narrowly—to only militaries of the EU member states or broadly—to the U.S. military operating in EU member states.[179] Peter P. Swire, formerly the U.S. government's Chief Counselor for Privacy, and co-author, Robert E. Litan, have contended in their study of the directive that the public security and defense exception would apply to the U.S. military operating in EU member states. Also, they believe a strong argument can be made that Article 25's "adequacy" requirement would be satisfied because the Privacy Act protects the privacy interests of U.S. nationals in the U.S. military and the federal government.

Thus, the EU would determine that the Privacy Act provides an "adequate level of protection." However, as Swire and Litan point out, the Privacy Act does not extend to foreign nationals. "Difficulties could arise, therefore, with records kept by the U.S. government about employees or other persons who are foreign nationals, such as when their employment or medical records are trans-

---

[178]See Paragraph (9) of Commission Decision Pursuant to Directive 05/46/EC of the European Parliament and of the Council on the adequacy of the protection provided by the Safe Harbor Privacy Principles and related Frequently Asked Questions, issued by the U.S. Department of Commerce, available at http://europa.eu.int/comm/internal_market/en/media/dataprot/news/decision.pdf.

[179]Article 3(2) of the directive reads in pertinent part:

> This Directive shall not apply to the processing of personal data: in the course of an activity which falls outside the scope of Community law, such as those provided for by Titles V and VI of the Treaty on European Union and in any case to processing operations concerning public security, defense, State security (including the economic well-being of the State when the processing operation relates to State security matters) and the activities of the State in areas of criminal law.

See *Directive*, available at http://www2.echo.lu/legal/en/dataprot/directiv/chap1.html#HD_NM_29.

ferred back to Washington" (Swire and Litan, 1998, p. 129). At any rate Swire and Litan do not believe the EU Commission will want to target the U.S. government for an early enforcement action because of the special diplomatic and legal problems such an action would raise (Swire and Litan, 1998, p. 129).

The directive's other exemptions may come into play. For example, Article 13(1) permits EU states to adopt legislative measures to restrict the scope of certain of the directive's obligations and rights, provided the restriction constitutes a necessary measure to safeguard national security, defense, and public security, among others.[180]

Moreover, Article 26(1) provides for derogation, or the partial revocation of a law, from Article 25. Specifically, EU member states "shall provide that a transfer or a set of transfers of personal data to a third country which does not ensure an adequate level of protection within the meaning of Article 25(2) may take place on condition that: . . . (d) the transfer is necessary or legally required on important public interest grounds."[181]

In principle, it may be argued that the exemptions provided for in Article 13 and 26(1) refer to national security reasons of the EU member states only, not the national security reasons of a foreign country, such as the United States. Nevertheless, the plain language of the Article 13(b) exception is "defence." Hence, a broad interpretation of defense cannot be ruled out. Similarly, Article 26(1)(d) establishes a derogation on the condition that "the transfer is necessary or legally required on important public interest grounds." Arguably, the U.S. Army's presence in EU member states serves an important public interest. Unfortunately, there is not much official guidance or scholarly work on the eventual application of the national security exception to this or similar cases.

Most important, many international agreements are in force between the United States and EU member states where the United

---

[180]Obligations and rights provided for in Articles 6(1), 10, 11(1), 12, and 21 of the directive may be restricted. See *Article 13(1) of the Directive*, available at http://www2.echo.lu/legal/en/dataprot/directiv/chap2.html#HD_NM_34.

[181]See *Article 26(1)(d) of the Directive*, available at http://www2.echo.lu/legal/en/dataprot/directiv/chap4.html

States has a military presence. These agreements, some of them classified, would pertain to the original grants of rights for the U.S. military presence in the host nation.[182] NATO multilateral agreements are also in force. These international agreements may contain provisions for derogation of some or all of the directive's obligations for the U.S. Army as a data controller. In case they do contain such provisions, the case law of the European Court of Justice should be reviewed to assess the impact of these international agreements of the EU member states with regard to EU Community Law.

Although the directive and its implementation are too recent to allow full evaluation of the precise impact the directive will have on Army biometric applications, on balance it appears likely that Army biometric applications will qualify for exemption from the directive's requirements. Also, to the extent that some U.S. Army bases are considered "joint" bases between the United States and the host nation EU member state, or to the extent these U.S. Army bases serve the defense of the EU member state by virtue of the international agreements to which they subscribed (e.g., mainly NATO), it is reasonable to think that EU member states could eventually authorize at least some of the exceptions provided for in the directive either with regard to the processing of personal data within the U.S. bases or the transfer of human resources data to the United States.

It should be noted, however, that the extent and scope of the exemptions and restrictions provided for in the directive are matters within the competence of the individual member states. Consequently, the precise interpretation of the exemptions could differ from one member state to another.

The final means for compliance with the directive—seeking an adequacy determination from the EU—is likely not a viable option given the political sensitivities involved. In any event, based on the discussion above, the Army will likely have no need to resort to this option.

Although the application of the EU Data Protection Directive to U.S. Army biometric applications appears complicated and confusing, the Army should bear in mind that it has experienced institutional assets

---

[182]For example, the bilateral agreement with Italy is classified.

on whom it may call, including Army Judge Advocate General and DoD OGC as well as EUCOM (European Command), USAREUR (United States Army, Europe), SHAPE (Supreme Headquarters Allied Powers Europe (NATO)) and the U.S. Department of State, who have dealt with similar issues in the past. Moreover, regardless of what the Army does with biometric applications and where it does it, the directive's applicability to the U.S. Army operating in EU member nations will eventually have to be decided because the Army is a huge collector of personal data in Europe and the directive defines personal data broadly and levies many requirements on the data collector.

Other International Law Concerns

As explained in the EU subsection above, when the Army operates in an overseas environment, there are some situations in which it is desirable for the Army to comply with foreign laws and some situations in which it is not desirable to do so. For example, the Italian government recently tried to require the U.S. military in Italy to accept Italian occupational safety and health laws. Other host nations attempt to force the U.S. military to accept their labor or environmental laws.

Just as with EU member states, the United States has entered into many bilateral agreements with other host nations where it has a military presence. For example, the United States and Japan have many such bilateral agreements (U.S. Forces Japan, 2000). These bilateral agreements can provide guidance for the Army as it plans biometric applications oversees. For example, many of these agreements give the Army great discretion in force protection and operational matters.[183] Again, once it determines exactly what type of biometric application it wants to require in an overseas location, the Army needs to look for specific legal guidance from its institutional assets to determine how it should proceed.

[183]See, e.g., Article III, Section 1, of the Agreed Minute to the Treaty of Mutual Cooperation and Security, dated January 19, 1960, providing that "[w]ithin the facilities and areas [Japan has permitted the United States to use], the United States may take *all the measures necessary for their establishment, operation, safeguarding and control*" (emphasis added), available at http://www.yokota.af.mil/usfj/Treaty2.htm.

CONCLUSION

This review has attempted to address legal concerns raised by Army use of biometrics. While not discussing every conceivable legal objection, this review was intended to provide the Army leadership with a useful starting point for legal analysis as it embarks on biometric applications. This review has also explained that while Army biometric applications raise legal concerns, these concerns about a new technology can be accommodated by the Army's and DoD's many institutional assets responsible for privacy issues. From the legal perspective, Army use of biometrics gets a tentative "Good to Go" for U.S.-based applications, provided it follows the statutory, administrative, and constitutional requirements discussed in this review.

In the international setting, the Army needs to be aware of the requirements of the EU Data Protection Directive and its impact on the U.S. government. Although the Army's use of biometrics will likely comply with the directive through one of the directive's exceptions, the Army must pay close attention to the way these exceptions are interpreted to avoid any difficulties. Moreover, any U.S. Army biometric application operating in a foreign nation must be examined from the international law perspective with the relevant bilateral agreements authorizing the U.S. Army's presence in the foreign nation as a starting point.

Before implementing any biometric application, the Army must undertake a thorough legal analysis of exactly what it wants to do and where it wants to do it. In this way, the Army will be much less likely to run afoul of the Privacy Act and similar statutory and administrative requirements, as well as the Constitution.

Appendix D

BIOMETRIC CONSORTIUM

The Biometric Consortium (BC) serves as the U.S. government's focal point for research, development, test, evaluation, and application of biometric-based personal identification/verification technology.[1] The BC's 700 members include federal, state, and local government officials; biometric industry representatives; academics; and representatives from related technologies. Jeffrey S. Dunn of the NSA and Fernando Podio of the National Institute of Standards and Technology (NIST) currently co-chair the BC.

The BC plays an important role in educating policymakers and the public about biometrics. For example, the BC chairs have testified before Congress and briefed senior Executive Branch officials on biometrics.[2] In addition, they have spoken at many leading government, industry, and academic conferences.

The BC sponsors conferences and other meetings as required. It publishes proceedings from its conferences and hosts a Web site. Among its responsibilities, the BC addresses legal and ethical issues surrounding biometrics. It also advises and assists member agencies concerning biometric technologies as well as the selection and application of biometric devices.[3]

---

[1]The BC Web site is available at http://www.biometrics.org/.

[2]See, e.g., Dunn (1998).

[3]BC Co-chairs Dunn and Podio both generously assisted RAND's research efforts for this project.

Chartered as a Working Group on December 7, 1995, the BC answers to the U.S. Security Policy Board through its Facilities Protection Committee. The Security Policy Board consists of the Director of Central Intelligence, Deputy Secretary of Defense, Vice Chairman of the Joint Chiefs of Staff, Deputy Secretary of State, Under Secretary of Energy, Deputy Secretary of Commerce, Deputy Attorney General, one Deputy Secretary from another nondefense-related agency, and one representative from OMB and the NSC staff.

In sum, the BC is one of the federal government's leading institutional assets in the field of biometric technologies.

Appendix E

INDIVIDUALS INTERVIEWED

Allen-Castellitto, Anita	Professor of Law, University of Pennsylvania School of Law
Alpert, Sherry	IRS
Baker, Stewart A.	Steptoe & Johnson
Belair, Robert R.	Mullenholz, Brimsek & Belair
Blumenthal, Marjory S.	National Research Council
Boesman, William C.	Congressional Research Service
Bowman, Erik	Identicator Technology
Brown, Linda T.	Infineon Technologies Corporation
Carter, Richard	American Association of Motor Vehicle Administrators
Cavoukian, Ann	Information and Privacy Commissioner, Ontario, Canada
Ciriaco, May Catherine	SSS-ID Project, Republic of the Philippines
Congour, David	Technical Security Division, U.S. Secret Service
Crawford, Susan	Wilmer, Cutler & Pickering
Davies, Simon	Privacy International
Di Dio, Arthur S.	Arent, Fox, Kintner, Plotkin & Kahn, PLLC
Dunn, Jeffrey	Co-Chair, Biometric Consortium
Emmanuel, Ezekiel	National Institutes of Health
Ford, Sheila	Office of the Secretary of Defense
Goldberg, Steve	Professor of Law, Georgetown University Law Center
Grippo, Gary	Program Manager for Electronic Money, U.S. Treasury
Higgins, Peter T.	Higgins & Associates, International
Hirst, Peter	London School of Economics

Hooghiemstra, Theo	Dutch Data Protection Authority
Howe, Randy	Uniformed Services University of Health Sciences
Jelinski, Steve	INS, Los Angeles International Airport
Kaneshiro, Julie	NIH—Office of the Director
Kelman, Alistair	LSE Enterprise
Kowalczyk, Jay	MITRE
Mahoney, Michael	Lone Star Technology Department, Texas Department of Social Services
Mansfield, Anthony	National Physical Laboratory, UK
McCreless, Kenneth	Deputy Marshal (Security), U.S. Supreme Court
Megna, Joe	Recognition Systems
Meslin, Eric	Executive Director, National Bioethics Advisory Commission
Mintie, Dave	Department of Social Services, State of Connecticut
Morgan, David P.	STS International
Nanavati, Raj	International Biometric Group, LLC
Nasrallah, F.P.	Assistant Professor of Ophthalmology, George Washington University
Negin, Michael	SENSAR
Noble, Kirsten Rudolph	Visionics Corp.
Norton, Richard E.	International Biometric Industries Association
Nunno, Richard	Congressional Research Service
Phillips, Jonathon	NIST
Podio, Fernando	Co-Chair, Biometric Consortium
Schellberg, Timothy M.	Smith Alling Lane, P.S.
Slagle, Geoffrey	American Association of Motor Vehicle Administrators
Smith, David	The Data Protection Registrar, UK
Spikes, Brent	Texas Department of Human Services
Steinfield, Lauren	Office of Management and Budget
Steinhardt, Barry	ACLU
Sure, Patrick	SAGEM SA, Paris—LA DEFENSE
Swire, Peter P.	Chief Counselor for Privacy, The White House
Szilagyi, Catherine A.	Steptoe & Johnson
Wales, Charlotte	MITRE
Warmack, Richard	U.S. Naval Criminal Investigative Service
Wayman, James L.	National Biometric Test Center, San Jose State University

Weedn, Victor W.	Carnegie Mellon University
Weete, John D.	West Virginia University
Weiss, Peter	Office of Management and Budget
Wheeler, John	SENSAR
Wilhelm, Catherine	Illinois Department of Human Services
Wilkinson, Harry	SecurCom
Wing, Bradford J.	INS
Yura, Michael	West Virginia University

BIBLIOGRAPHY

Aaron, Ambassador David L., cover letter to U.S, organizations requesting comments on the newly posted draft documents, November 15, 1999, available at http://www.ita.doc.gov/td/ecom/aaronmemo1199.htm.

Agre, Philip E., and Marc Rotenberg, *Technology and Privacy: The New Landscape*, Cambridge, Mass.: MIT, 1998.

Alabama Lawsuit Challenging the State Requirement for Social Security Numbers from Driver License Applicants, available at http://www.networkusa.org/fingerprint/page1b/fp-al-suit-aug99-update.htm, 2000.

Alderman, Ellen, and Caroline Kennedy, *The Right to Privacy*, New York: Vintage Books, 1995.

Allen, Anita, "Legal Issues in Nonvoluntary Prenatal Screening in AIDS," in Ruth R. Faden et al., eds., *Women and the Next Generation: Towards a Mutually Acceptable Public Policy for HIV Testing of Pregnant Women and Newborns*, 1991, p. 175.

American Law Reports, "Validity of Statute, Ordinance, or Regulation Requiring Fingerprints of Those Engaged in Specific Occupations," Vol. 41, 1972 (1999 supplemental), p. 732.

Archer, Charles, W., Assistant Director, Criminal Justice Information Services Division, FBI, on Criminal Information Services Division, statement for the record before the Subcommittee on Immigration, Committee on Judiciary, U.S. Senate, May 1, 1997, available at http://www.fbi.gov/pressrm/congress/congress97/archer.htm.

Armed Forces Repository of Specimen Samples for the Identification of Remains, A0040-57a, available at http://www.defenselink.mil/privacy/notices/army/A0040-57a-DASG.html, 2000.

Army Regulation 380-67, *Army Personnel Security Program*, September 9, 1988.

Army Regulation 340-21, *Army Privacy Program*, July 5, 1985.

Army Regulation 600-20, 5-6 *Accommodating Religious Practices*, July 15, 1999.

Baker, Stewart A., and Paul R. Hurst, *The Limits of Trust: Cryptography, Governments, and Electronic Commerce*, Boston: Kluwer Law International, 1998.

Barrett, William A., *Iriscan Evaluation*, National Biometric Test Center, San Jose State University, March 10, 1999.

_____, *Image Capture System*, Computer Information and Systems Engineering Department, San Jose State University, July 21, 1999.

_____, *Some Observations on the Cumulative Binomial Probability Distribution*, Computer Information and Systems Engineering Department, San Jose State University.

Belsie, Laurent, "Tech Trends Coming Soon: ATMs That Recognize Your Eyes," *Christian Science Monitor*, December 2, 1997.

Bevier, Lillian R., "Information About Individuals in the Hands of Government: Some Reflections on Mechanisms for Privacy Protection," *William & Mary Bill of Rights Journal*, Vol. 4, 1995.

Bork, Robert H., *The Tempting of America: The Political Seduction of the Law*, New York: Free Press, 1990.

Borking, J. J., and B. M. A. van Eck, *Intelligent Software Agency and Privacy*, Registratiekamer, The Hague, the Netherlands, January 1999.

Brandeis, Louis D., and Samuel D. Warren, "The Right to Privacy," *Harvard Law Review*, Vol. 4, 1890.

Brin, David, *The Transparent Society: Will Technology Force Us to Choose Between Privacy and Freedom?* Reading, Mass.: Perseus Books, 1998.

Buckeley, William M., "Your Face or Mine? Ask a Computer, Feature-Recognition Systems Match Facial 'Landmarks' to Determine Positive ID," *Wall Street Journal*, December 7, 1999, p. B4.

Byrd, Senator Robert C., statement, *Congressional Record–Senate*, Vol. 145, June 8, 1999, pp. 6648–6651.

Cate, Fred H., *Privacy in the Information Age*, Washington, D.C.: The Brookings Institution, 1997.

Chen, Harold, *Medical Genetics Handbook*, St. Louis: W. H. Green, 1998.

CJIS, Programs Support Section, "Use of Civil Fingerprint Records for Investigative Purposes," memorandum, March 18, 1998.

Clarke, Roger, "Human Identification in Information Systems: Management Challenges and Public Policy Issues," *Information, Technology, and People*, December 1994.

Clinton, President William J., "State of the Union Address," January 27, 2000.

CNN, "Popular Software Secretly Sends Music Preferences," November 1, 1999.

_____, "RealNetworks Apologizes for Privacy Breach," November 2, 1999.

Cavoukian, Ann, and Don Tapscott, *Who Knows: Safeguarding Your Privacy in A Networked World*, New York: McGraw-Hill, 1996.

Connecticut State Department of Social Services (DSS), Digital Imaging: Connecticut's Digital Imaging Project, available at http://www.dss.state.ct.us/digital/project.htm, 2000.

_____, Digital Imaging Program Fact Sheet, January 1996, available at http://www.dss.state.ct.us/pubs/difacts.pdf.

_____, Office of Administrative Hearings and Appeals, *Notice of Decision*, November 10, 1998.

_____, Office of Administrative Hearings and Appeals, *Order to Vacate Hearing Decision*, May 29, 1999.

D'Agnese, Joseph, "The Technology Is Here: Why Can't You Buy One?" *Discover*, Vol. 20, No. 9, September 1999.

Davies, Simon G., "Touching Big Brother: How Biometric Technology Will Fuse Flesh and Machine," 1994, available at http://www.privacy.org/pi/reports/biometric.html.

Department of Commerce, *Safe Harbor Workbook,* available at http://www.export.gpv/safeharbor/SafeHarborWorkbook.htm, 2000.

Department of Defense Directive 5400.11, *Department of Defense Privacy Program,* June 9, 1982.

Department of Defense Directive 5400.11-R, *Department of Defense Privacy Program,* August 31, 1982.

Department of Defense, Privacy Act Systems of Records Notices, available at http://www.defenselink.mil/privacy/notices/, 2000a.

_____, Defense Privacy Board Opinion 34, *Definition of 'Order of a Court of Competent Jurisdiction,'* available at http://www.defenselink.mil/privacy/opinions/op0034.html, 2000b.

_____, Privacy Office Mission and Functions, available at http://www.defenselink.mil/privacy/mis-func.html, 2000c.

_____, *Biennial Report for Calendar Years 1996–1997,* available at http://defenselink.mil/privacy/pdfdocs/18jun98ltr2omb.pdf, 2000d.

Department of Justice, Office of Information and Privacy and Office of Management and Budget, Office of Information and Regulatory Affairs, *Overview of the Privacy Act of 1974,* available at http://www.usdoj.gov/04foia/1974intro.htm, 1998.

_____, Bureau of Justice Statistics, "Legal and Policy Issues Related to Biometric Identification Technologies," April 1990, pp. 48–52.

_____, Office of Legal Education, Executive Office for United States Attorneys, *Privacy Act,* November 1999.

Department of the Army, "Privacy Act Systems of Records Notices," available at http://www.defenselink.mil/privacy/notice/army/, 2000.

Directive 94/46/EC of the European Parliament and of the Council of 24 October 1995 on the Protection of Individuals with regard to the Processing of Personal Data and on the Free Movement of Such Data, available at http://www2.echo.lu/legal/en/dataprot/directiv/directiv.html, 2000.

Dunn, Jeffrey S., "Biometrics and the Future of Money," testimony before the Subcommittee on Domestic and International Monetary Policy, Committee on Banking and Financial Services, U.S. House of Representatives, May 20, 1998, available at http://www.house.gov/banking/52098jd.htm.

Dupont, Daniel, "GAO: Army's Plans for First Digitized Division Rife with Uncertainty," *Defense Information and Electronics Report,* July 30, 1999.

Economist, The, "Living in the Golden Fishbowl," December 18, 1999.

Electronic Privacy Information Center, *New White House Computer Surveillance Plan Would Pose Unpredicted Threat To Privacy,* August 20, 1999.

_____, *Privacy & Human Rights: An International Survey of Privacy Laws and Developments,* 1999.

_____, *FCC Approval of FBI Wiretap Standards Threatens Communications Privacy,* August 27, 1999.

Etzioni, Amitai, *The Limits of Privacy,* New York: Basic Books, 1999.

Executive Order 10450, § 3 (a), *Security Requirements for Government Employees,* dated April 27, 1953, available at http://www.nara.gov/fedreg/eos/e10450.html, 2000.

Executive Order 12333, dated December 4, 1981, 46 Fed. Reg. 59941.

FBI, press release, August 10, 1999, available at http://www.fbi.gov/pressrm/pressrel/pressrel99/iafis.htm.

_____, Office of General Counsel, *Using Civil Fingerprint Records For Criminal Justice Purposes,* opinion, August 23, 1995.

Finch, Lewis C., Acting Under Secretary of Defense (Personnel and Readiness), "Fingerprint Capture Policy," memorandum, July 15, 1997.

Fried, Charles, *An Anatomy of Values,* Cambridge, Mass.: Harvard University Press, 1970.

GAO, *Social Security: Government and Commercial Use of the Social Security Number Is Widespread*, Washington, D.C.: GAO, 1999, GAO/HEHS-99-28.

GAO Report, *Electronic Benefits Transfer: Use of Biometrics to Deter Fraud in the Nationwide EBT Program,* September 9, 1995.

_____, *Illegal Immigration: Southwest Border Strategy Results Inconclusive; More Evaluation Needed,* December 12, 1997.

Garfinkel, Simson, *Database Nation: The Death of Privacy in the 21st Century*, Sebastopol, Calif.: O'Reilly and Associates, 2000.

Gavison, Ruth, "Privacy and the Limits of Law," *Yale Law Journal,* Vol. 89, 1980.

Gellman, Robert, "Feel Like a Number," August 27, 1998, available at http://intellectualcapital.com/issues/98/0827/iccon.asp.

General Services Administration, *Smart Access Common ID Card: Final Requirements Document,* July 2, 1999.

Gerety, Tom, "Redefining Privacy," *Harvard Civil Rights-Civil Liberties Law Review*, Vol. 12, 1977.

Gillert, Douglas J., "Who Are You? DNA Registry Knows," *American Forces Press Service,* July 1998.

Goldberg, Steve, *Culture Clash: Law and Science in America,* New York: New York University Press, 1994.

Government Computer News, "Visionics Corp. Changes the Face of PC Security," November 24, 1997.

Gugliotta, Guy, "Bar Codes for the Body Make It to the Market," *Washington Post,* June 21, 1999, p. A1.

Hansell, Saul, "Use of Recognition Technology Grows in Everyday Transactions, *New York Times,* August 27, 1997.

Harreld, Heather, "Biometrics Points to Greater Security: Agencies Eye the Latest Devices to Verify IDs," *Federal Computer Week,* July 19, 1999.

Hawkes, Peter, and Stewart Hefferman, "Biometrics: Understanding the Business Issues in Just One Day," presentation, SJB Biometrics '99 Workshop, November 9, 1999.

Hes, R., T. F. M. Hooghiemstra, J. J. Borking, *At Face Value: On Biometrical Identification and Privacy,* 2nd Edition, Registratiekamer, The Hague, the Netherlands, November 1999.

Hes R., and J. J. Borking, *Privacy-Enhancing Technologies: The Path to Anonymity,* Registratiekamer, The Hague, the Netherlands, November 1998.

Hixson Richard F., *Privacy in a Public Society: Human Rights in Conflict,* New York: Oxford University Press, 1987.

Hornack, L. A., *The Center for Identification Technology Research (CITeR),* Department of Computer Science and Electrical Engineering, West Virginia University, Huntington, W. Va.

"Identification Division Records System Notice," printed in *Federal Register,* Reg. 49174, Vol. 55, No. 227, November 26, 1990.

Illinois State Department of Human Services (DHS), Bureau of Program Design and Evaluation, *Evaluation Report: Biometric Identification Demonstrations,* Springfield, Ill.: State of Illinois, 1997.

Jacobson, Louis, "Playing the Identity Card," *National Journal,* Vol. 31, No. 12, March 20, 1999.

Jain, Anil, Ruud Bolle, and Sharath Pankanti, *Biometrics: Personal Identification in Networked Society,* Boston: Kluwer Academic Publishers, 1998.

Katsh, M. Ethan, *Law in a Digital World,* Oxford, U.K.: Oxford University Press, 1995.

LaFave et al., *Criminal Procedure,* Vol. 2, St. Paul, Minn.: West Publishing, 1999, § 3.2 (g).

Lessig, Lawrence, *Code and Other Laws of Cyberspace,* New York: Basic Books, 1999.

LeDuc, Daniel, "'Smart Gun' Mandate Urged by Md. Panel," *Washington Post,* November 10, 1999.

LeDuc, Daniel, and Craig Whitlock, "Guns Top Legislative Agenda in Maryland," *Washington Post,* January 25, 2000, p. B1.

Lynch, Shannon, "Maryland in Fight to Keep Medical Records Private," *Frederick News-Post,* October 21, 1999.

Messmer, Ellen, "Pentagon Gets 'Smart'; Military smart cards will access nets, encrypt data," *Network World,* September 20, 1999.

Miller, John J., and Stephen Moore, *A National ID System: Big Brother's Solution to Illegal Immigration,* policy analysis No. 237, Cato Institute, September 7, 1995.

Mintie, David, ed., *Biometrics in Human Services, User Group Newsletter,* various issues, available at http://www.dss.state.ct.us/digital/faq/dihsug.htm.

Moore, Lloyd T., "Biometrics and Smart Cards Go to Boot Camp," *Business Systems Magazine,* September 1998, available at http://www.corrypub.com/bsm/articles/Sept.98/099814.cfm, 2000.

Moore, Stephen, "A National Identification System," testimony before the U.S. House of Representatives Subcommittee on Immigration and Claims Judiciary Committee, 1997, available at http://www.cato.org/testimony/ct-sm051397.htm.

Moskowitz, Robert, "Are Biometrics Too Good?" *Network Computing,* January 25, 1999.

Murphy, Richard, "Property Rights in Personal Information: An Economic Defense of Privacy," *Georgetown Law Journal,* Vol. 84, 1996.

Newton, Elaine M., and David Rubenson, interview with James L. Wayman, October 25, 1999.

Newton, Elaine M., and Katharine W. Webb, interview with Tony Mansfield, National Physical Laboratory, London, England, November 9, 1999.

Noguchi, Yuki, "Techway; Technology for a Swift Eye-D; A Falls Church Entrepreneur Thinks Iris Scans Will Replace Tickets for Travel and Sports Events," *Washington Post,* November 8, 1999.

Nuger, Kenneth P., and James L. Wayman, *Reconciling Government Use of Biometric Technologies With Due Process and Individual Privacy,* National Biometric Test Center, San Jose State University, Calif.

Office of Management and Budget, *Management of Federal Information Resources;* notice, March 5, 1999.

_____, *Guidelines,* 40 Fed. Reg. At 28, 948, 1975, and 52 Fed. Reg. 12,990, 1987.

O'Harrow, Robert, "Firm Changes to Plan to Acquire Photos; Driver's Pictures Ignited Privacy Furor," *Washington Post,* November 12, 1999.

_____, "A Hidden Toll on Free Calls: Lost Privacy; Not Even Unlisted Numbers Protected from Marketers," *Washington Post,* December 19, 1999.

Parker, Richard B., "A Definition of Privacy," *Rutgers Law Review,* Vol. 27, 1974.

Piller, Charles, et al., "Super Day for Big Brother," *Los Angeles Times,* February 2, 2001.

The Privacy Act of 1974, codified at 5 U.S.C. § 552a.

Privacy and American Business, *A Comprehensive Report and Information Service,* Vol. 6, No. 5, September/October 1999.

Proceedings of the IEEE, "Error-Rate Equations for the General Biometric System," March 1999.

_____, "Special Issue on Automated Biometric Systems," Vol. 85, No. 9, September 1997.

Ratha, N. K., and Ruud Bolle, "Smart Card Based Authentication," in Anil K. Jain and Ruud Bolle, eds., *Biometrics: Personal Identification in a Networked Society,* Boston: Kluwer Academic Publishers, 1999, Chapter 18.

Roch, Michael P., "Filling the Void of Data Protection in the United States Following the European Example," *Santa Clara Computer & High Technology Law Journal,* Vol. 12, 1986.

Rogers, William, ed., *Biometric Digest,* various issues.

Schwartz, Paul M., "Privacy and Participation: Personal Information and Public Sector Regulation in the United States," *Iowa Law Review,* Vol. 80, 1995.

Scott, W., and M. Jarnigan, *Treatise upon the Law of Telegraphs,* Appendix, New York: Little Brown, and Co., 1868, pp. 457–507.

700 Club, The, "Biometrics—Chipping Away Your Rights," fact sheet, October 9, 1995.

Sinatra, Amy, "Building Safer Guns," ABC News, available at http://abcnews.go.com/sections/us/DailyNews/guns_safety.html, December 17, 2000.

SJB, *The Biometric Report,* 1999.

Slevin, Peter, "Police Video Cameras Taped Football Fans," *Washington Post,* February 1, 2001.

Stamper, Chris, "Fighting the Fingerprints," CNN, July 24, 1997, available at http://www.cnn.com/tech/9707/24/netly.news/index.html, 2000.

Sovereign Citizens Against Numbering (SCAN) Web site, "Fight the Fingerprint!" available at http://www.networkusa.org/fingerprint.shtml, 2000.

Strauss, Peter L., et al., *Gellhorn & Byse's Administrative Law: Cases & Comments,* 1995.

Swire, Peter P., and Robert E. Litan, *None of Your Business: World Data Flows, Electronic Commerce, and the European Privacy Directive,* Washington, D.C.: Brookings Institution, 1998.

Tomko, George, "Biometrics as a Privacy-Enhancing Technology: Friend or Foe of Privacy?" Privacy Laws and Business Ninth Privacy Commissioners'/Data Protection Authorities Workshop, September 15, 1998.

Tuchman, Gary, "New York to Expand DNA Testing of Convicts," CNN, October 20, 1999.

Turkington, Richard C., and Anita L. Allen, *Privacy Law: Cases and Materials,* St. Paul, Minn.: West Publishing, 1999.

Uniform Policy Manual Index § 8540.70, "TFA Non-Financial Requirements, Digital Imaging Process."

U.S. Forces Japan, reference library, available at http://www.yokota.af.mil/usfj/LIBRARY.htm, 2000.

U.S. Senate, Amendment No. 594, Department of Defense Appropriations Act, S. 1112, 2000.

U.S. Senate, S. Report No. 93-1183, reprinted in 1974 U.S. Code Cong. and Admin. News 6916.

U.S. Social Security Administration, Social Security Number Chronology, 1998, available at http://www.ssa.gov/history/ssnchron.htm.

"Validity of Statute, Ordinance, or Regulation Requiring Fingerprints of Those Engaged in Specific Occupations," *American Law Reports,* 3rd Edition, Vol. 41, 1972, Supplement 1999.

Van Kralingen, Robert, et al, "Using Your Body as a Key," *Legal Aspects of Biometrics,* November 1997.

Wampler, Stephen, *Lab Scientists Deliver First Battery-Operated, Portable DNA Analysis System to the U.S. Army,* Lawrence Livermore National Laboratory, November 7, 1996.

Warren, Earl, "The Bill of Rights and the Military," *New York University Law Review*, Vol. 37, 1962.

Warren, Samuel D., and Louis D. Brandeis, The Right to Privacy, *Harvard Law Review*, Vol. 4, 1890, p. 193.

Wayman, James L., *Confidence Interval and Test Size Estimation for Biometric Data*, National Biometric Test Center, San Jose State University, Calif., October 1995.

_____, *The State of Biometrics: Standards, Alliances and Applications*, National Biometric Test Center, San Jose State University, Calif., 1998.

_____, *Multi-Finger Penetration Rate and ROC Variability For Automatic Fingerprint Identification Systems*, National Biometric Test Center, San Jose State University, Calif., September 1999a.

_____, *When Bad Science Leads to Good Law: The Disturbing Irony of the Daubert Hearing in the Case of U.S. v. Byron C. Mitchell*, National Biometric Test Center, San Jose State University, Calif., October 1999b.

_____, "Technical Testing and Evaluation of Biometric Identification Devices," in Anil K. Jain and Ruud Bolle, eds., *Biometrics: Personal Identification in a Networked Society*, Boston: Kluwer Academic Publishers, 1999c, Chapter 17.

_____, *Biometric Technology: Testing, Evaluation, Results*, National Biometric Test Center, San Jose State University, Calif., 1999d.

_____, *Technical Testing and Evaluation of Biometric Identification Devices*, National Biometric Test Center, San Jose State University, Calif., 1999e.

_____, *Fundamentals of Biometric Technologies*, National Biometric Test Center, San Jose State University., Calif., 1999f.

_____, *Biometrics and the Future of Money*, testimony before the Subcommittee on Domestic and International Monetary Policy, Committee on Banking and Financial Services, U.S. House of Representatives, May 20, 1998, available at http://www.house.gov/banking/52098jlw.htm.

_____, "Federal Biometric Technology Legislation," *Proceedings of the IEEE,* February 2000, available at http://computer.org/computer/articles/February/coverfeature200_1.htm.

Weedn, Victor Walter, "Stored Biological Specimens for Military Identification: The Department of Defense DNA Registry," in Robert F. Weir, ed., *Stored Tissue Samples: Ethical, Legal, and Public Policy Implications,* Iowa City, Iowa: University of Iowa Press, 1998.

Weiss, Peter N., *Access America—Fulfilling the Vision of Electronic Service Delivery,* Information Policy and Technology, Office of Management and Budget.

Westin, Alan, *Privacy and Freedom,* New York, Atheneum, 1967.

Westin, Alan, *Public Attitudes Toward the Use of Finger Imaging Technology for Personal Identification in Commercial and Government Programs,* National Opinion Survey, conducted by Opinion Research Corporation, August 1996.

Woodward, John D., Jr., *Biometrics and the Future of Money,* testimony before the Subcommittee on Domestic and International Monetary Policy, Committee on Banking and Financial Services, U.S. House of Representatives, May 20, 1998, available at http://www.house.gov/banking/52098jdw.htm.

_____, "Biometric Scanning, Law and Policy: Identifying the Concerns—Drafting the Biometric Blueprint," 59 *University of Pittsburgh Law Review,* Vol. 59, 1997a.

_____, "Biometrics: Privacy's Foe or Privacy's Friend?" *Proceedings of the IEEE,* Vol. 85, September 1997b.

_____, "And Now the Good Side of Facial Profiling," *Washington, Post,* February 4, 2001.

Woodward, John D., Jr., and Gary Roethenbaugh, "Fact Sheet on the European Union Privacy Directive," available at http://www.dss.state.ct.us/digital/eupriv.html, 2000.

Woodward, John D., Jr., and Katie Smythe, interview with Geoffrey Slagle, Standards Program Director, AAMVAnet, March 28, 2000.